STP 1109

Modern Instrumental Methods of Elemental Analysis of Petroleum Products and Lubricants

R. A. Nadkarni, editor

ASTM Publication Code Number (PCN):
04-011090-12

ASTM
1916 Race Street
Philadelphia, PA 19106

Library of Congress Cataloging-in-Publication Data

Analysis of petroleum products and lubricants / R. A. Nadkarni, editor.
"STP 1109."
"The ASTM Symposium on Modern Instrumental Methods of Elemental Analysis of Petroleum Products and Lubricants was held in New Orleans, Louisiana, on 11–13 December 1989"—Foreword.
Includes bibliographical references and index.
ISBN 0-8031-1416-8
1. Petroleum products—Analysis—Congresses. 2. Lubrication and lubricants—Analysis—Congresses. I. Nadkarni, R. A. II. ASTM Symposium on Modern Instrumental Methods of Elemental Analysis of Petroleum Products and Lubricants (1989: New Orleans, La.) III. American Society for Testing and Materials.
TP691.A68 1991
665.5—dc20 91-10488
CIP

Peer Review Policy

Each paper published in this volume was evaluated by three peer reviewers. The authors addressed all of the reviewer's comments to the satisfaction of both the technical editor(s) and the ASTM Committee on Publications.

The quality of the papers in this publication reflects not only the obvious efforts of the authors and the technical editor(s), but also the work of these peer reviewers. The ASTM Committee on Publications acknowledges with appreciation their dedication and contribution to time and effort on behalf of ASTM.

Printed in Baltimore, MD
May 1991

Foreword

The ASTM Symposium on Modern Instrumental Methods of Elemental Analysis of Petroleum Products and Lubricants was held in New Orleans, Louisiana, on 11–13 December 1989. The event was sponsored by ASTM Committee D–2 on Petroleum Products and Lubricants. R. A. Nadkarni, Exxon Chemical Company, served as symposium chairman and has also edited this publication.

Contents

Overview

The last two decades have seen explosive growth in the development and use of analytical instrumentation in all areas of science. The field of petroleum and fuel science has certainly been no exception to this trend. In a commercial or industrial laboratory nowadays, one will rarely find classical wet chemistry techniques being used for elemental analysis. These are being increasingly replaced by modern instrumental techniques such as atomic absorption (AA) spectrometry, inductively coupled plasma atomic emission spectrometry and mass spectrometry, X-ray fluorescence (XRF), ion chromatography, and so on. Even a reasonably modern and universally used technique such as atomic absorption spectrometry is widely being replaced by a newer and better technique, inductively coupled plasma atomic emission spectrometry (ICPAES).

In spite of the advances in analytical technology, many product specifications continue to be written with older test methods specified for analysis, even though very few, if any, laboratories may be using these methods (e.g., ASTM D 811). Hence a symposium was organized by ASTM Committee D-2 on Petroleum Products and Lubricants and its Subcommittee D02.03 on Elemental Analysis to provide analysts and marketers with information on state-of-the-art methodology for the elemental analysis of petroleum products and lubricants. The symposium was held in New Orleans, Louisiana, on 11–13 December 1989 and was attended by over 150 people from nine countries. Twenty papers were presented at the symposium, from which thirteen are included in this ASTM Special Technical Publication. The papers have been arranged by analytical techniques, although some papers fall into more than one category.

The first two papers are comprehensive reviews by *Nadkarni*, the first describing the occurrence and significance of trace metals, the second presenting current analytical technology for elemental determination in petroleum-related materials. Trace metals in crude oils originate from the marine animal and vegetative source of crude oils. Many metals are present as porphyrin complexes and are often used as the biomarkers for the origin of crude oil. On the other hand, several metals may be added during refining to the petroleum or to the lubricating oils for diverse beneficial physical performance. Part I discusses comprehensively the role of various metals added to lubricants, the significance of trace elements found in used lubricating oils, and the various standard reference materials that can be used for quality assurance of petroleum analysis. ASTM publishes nearly 45 methods for the elemental analysis of petroleum products. Part II covers the diverse sample preparation schemes necessary for certain analytical techniques, and then reviews the principles and applications of each of the major techniques. The review covers atomic absorption spectrometry with and without flame, inductively coupled plasma atomic emission spectrometry, microelemental methods, neutron activation analysis, spark source mass spectrometry, X-ray fluorescence, and ion chromatography. Most of these techniques are fully matured and have been extensively documented in the literature. Yet no one technique can be ideal for all types of analyses. An analyst must make a pragmatic choice based on specific criteria for a specific matrix for a particular analysis.

Mackey et al. compare ICPAES and XRF as two alternative techniques for the determination of additive elements in lube oils, as practiced in their laboratory. The aim of the paper is to help the nonexpert get a better understanding of the advantages and drawbacks of these two techniques.

The next five papers deal with atomic spectroscopy analysis, both for crude and used oils and additives. *Gonzales and Lynch* describe an ICPAES method for the determination of 18 trace elements in crude oil. Optimum conditions for solvent selection, proper wavelength, and use of internal standard result in recoveries greater than 90% for the elements of interest.

Carter et al. describe an automated rapid multielement determination of wear metals in used oils using flame AA with a single aspiration per sample. This spectrometer can sequentially determine up to 24 elements per sample in a single run. The method appears much more powerful and streamlined than those using conventional AA instruments.

Lukas and Anderson describe three techniques that can be used to improve the ability of optical emission spectrometry to detect and quantify large wear particles in lubricating oil. These are the single spark, ashing rotrode, and acid digestion differential techniques. The technique of choice varies with the requirements and philosophy of the analytical laboratory.

In a similar comparison, *Nygaard et al.* compared 50 used and 20 new diesel engine oils for ten wear metals and four additive elements analyzed by ICPAES and rotating disk electrode spark emission spectrometry. Interestingly, the two groups of elements were found to respond differently to the two techniques. They were also affected by the sample viscosity. It is believed that rotrode is somewhat more effective in sampling particulates than is ICPAES.

The above papers attest to the continuing interest in the determination of wear metals in used oils using AAS or ICPAES. Currently, ASTM D02.03 is involved in completing an ICPAES test method for wear metals in used oils. A round-robin has been completed and the method is being ballotted.

The newly developed technique of inductively coupled plasma–mass spectrometry (ICP-MS) is the subject of a paper by *Williams*. She demonstrates the application of this technique to the determination of several environmentally important trace elements in recycled oils. The ICP-MS technique, even though widely used in geological and isotopic studies, lags far behind ICPAES in terms of its use in the petrochemicals industry. The cost and complexity of the instrument as well as the wide use of ICPAES have contributed to this situation.

A technique as important as atomic spectroscopy in petrochemicals analysis is X-ray fluorescence. Four papers describe the applications of this technique to matrices as varied as reforming catalysts to lubricating oils.

The paper by *McElroy and Mulhall* takes a quite different approach from what we normally think of as petrochemicals applications, although their application is a vital link in refining crude oils to petroleum. They describe an XRF method for assaying the platinum, rhenium, and iridium content of fresh reforming catalysts. Because millions of dollars ride on a precise assay, an interlaboratory agreement of ±0.5% relative standard deviation between the catalyst supplier and purchaser must be achievable by the analytical technique used.

Sieber et al. then describe the determination of lubricant additive elements by XRF. When XRF alone is insufficient to obtain accurate information, complementary techniques such as ICPAES, AAS, and wet chemistry are also used. Agreement between results by different techniques enables an analyst to check the accuracy of the methods and provide customers with reliable data.

In the third XRF paper, *Shay and Woodward* describe the use of energy dispersive XRF for the determination of porphyrin complexed metals in petroleum and residua. These metals (V, Ni, and Fe) generally have a negative impact on refining operations and hence must be quickly and accurately determined. The authors describe the approaches used to eliminate the matrix effects and to correct for spectral interferences.

The fourth paper on XRF applications is by *Wheeler*. This paper also describes the energy dispersive XRF technique applied to crude and lubricating oils. For precise and accurate results to be obtained, proper excitation conditions, sample volume, and support film, an adequate mathematical model for interelement corrections, and accurate standards are essential.

The final paper is by *Ohlson and Takahashi.* It describes a microcoulometric technique for the determination of trace amounts of halides, principally chloride, in hydrocarbons. Even though the combustion-coulometric technique is well established, certain types of heavy oils pose problems such as sluggish response due to refractory components, presence of heavy metals, and rapid aging of the quartz combustion tubes. This paper suggests a way to overcome these problems by changing the combustion tube design, adding copper oxide wire to the combustion outlet zone, and adding a post-combustion scrubber to remove deposits. There is certainly a need in the marketplace for an elegant and simple chlorine analyzer similar to those available for carbon, hydrogen, and nitrogen.

These papers provide the reader with the latest information on the state-of-the-art analytical instrumentation used for elemental analysis of petroleum products and lubricants. The major techniques used in this field are included here, although other techniques such as ion chromatography and microchemical analysis are missing.

I wish to acknowledge the prompt responses and cooperation received from the authors, reviewers, and the ASTM staff to make for a successful symposium and subsequent efficient publication of this volume. I also wish to thank the management of PARAMINS Technology Division, Exxon Chemical Company, for providing active support in my involvement in organizing the symposium and the subsequent book publication.

R. A. Nadkarni
Exxon Chemical Company
Linden, NJ 07036
Symposium Chairman & Editor

R. A. Nadkarni[1]

A Review of Modern Instrumental Methods of Elemental Analysis of Petroleum Related Material: Part I—Occurrence and Significance of Trace Metals in Petroleum and Lubricants

REFERENCE: Nadkarni, R. A., "**A Review of Modern Instrumental Methods of Elemental Analysis of Petroleum Related Material: Part I—Occurrence and Significance of Trace Metals in Petroleum and Lubricants,**" *Modern Instrumental Methods of Elemental Analysis of Petroleum Products and Lubricants, ASTM STP 1109,* R. A. Nadkarni, Ed., American Society for Testing and Materials, Philadelphia, 1991, pp. 5–18.

ABSTRACT: Almost all elements in the periodic table are found in petroleum products, varying from percent levels for C-H-S to parts per million and parts per billion levels for transition metals. The presence of elements in crude oil is ascribed to its marine animal and vegetative origin. Compared with two other fossil fuels — coal and oil shale — petroleum is a lesser source of environmental pollution. Extensive data on crude oils composition are available. Determination of S, N, V, Ni, and Fe form an integral part of crude assay analysis of oils carried out by all oil companies. Generally it is agreed that V, Ni, and Fe occur as metallo-porphyrins; Hg, Sb, and As as organometallic compounds; Mo and Ge as carboxylic acid salts; and silica and NaCl as colloidal minerals.

The significance of metals either naturally present in crude and gasoline or purposely added to petroleum products is described. Trace metals in used lubricating oils indicate the wear performance of an engine, and can be used as a preventive maintenance check. A number of sample preparation schemes are necessary depending on the analytical technique to be used. This can vary from procedures as simple as dilution for XRF or AAS, to more involved procedures such as oxygen bomb combustion for ion chromatography, or wet ashing for AAS or ICPAES. Frequently, robots are used to speed sample dilution for XRF or ICPAES.

For quality assurance of these analyses, standard reference materials (SRMs) are used. At least three sources of these materials are Alpha Resources, Leco Corporation, and the National Institute of Standards and Technology. SRMs for crude oil, diesel oil, fuel oil, kerosene, lube oil, paraffin oil, and petroleum coke are currently available.

KEY WORDS: petroleum analysis, origin of elements in oil, wear metals, additive elements, standard reference materials, sample preparation

The main constituents of petroleum products are, of course, hydrocarbons; however, all crude oils, petroleum products, lubricants, and additives contain metals and nonmetals, either naturally occurring or deliberately added. A large number of elements in the periodic table have been found in the petroleum products varying from percent levels of carbon,

[1] Analytical Process Leader, Chemical Analytical Laboratory, PARAMINS Technology Division, Exxon Chemical Company, Linden, NJ 07036.

hydrogen and sulfur, to ppm and ppb levels of transition metals. A vast body of literature exists regarding the occurrence, significance, and analysis of metals and nonmetals in crude oils and petroleum products.

Part I of this paper will describe the occurrence and significance of metals found in petroleum products, while Part II will discuss applications of individual analytical techniques in this area.[2]

Occurrence of Metals in Crude Oil

Generally, it is believed that the source of petroleum is of marine animal and vegetative life deposited with sediment in the coastal waters in prehistoric times. Due to bacterial action over a period of time sulfur, oxygen, nitrogen, and volatile compounds evolved out of this material, but not completely. The end-product was a mixture of hydrocarbons containing varying amounts of sulfur, oxygen and nitrogen, and trace amounts of metals and other elements.

Unlike the two other fossil fuels, coal and oil shale, petroleum is a lesser source of elemental pollution in the environment. Unlike coal, which is burned unaltered as an energy source, oil is extensively processed in the refinery to yield a wide range of usable products.

Generally, sulfur and nitrogen are the most abundant elements in crude oils except for carbon and hydrogen. Sulfur up to 6 wt% and nitrogen over 1 wt% have been reported. More than 40 other elements are present in crude oils at trace levels (<1000 ppm); most common are: vanadium (up to 1000 ppm), nickel (up to 100 ppm), and iron (up to 30 ppm); in lesser amounts As, Be, Cd, Hg, Se, Cu, Sb, Ba, Cr, Pb, Mn, Mo, Te, and Sn have been found. Extensive data on elemental composition of crude oils can be found in Ref *1*.

In general, the lowest metal containing U.S. crudes are from coastal and offshore fields of Texas and Louisiana. The highest metal containing U.S. crudes are found in California. A summary of worldwide crude oils is given in Table 1. Sulfur, nitrogen, vanadium, nickel, and iron are the most frequently determined elements in crude oils. There are probably millions of data points in this area, since these analyses are an integral part of "crude assay analysis" conducted by all oil companies. Most data, however, are considered proprietary. Some typical data on crudes from the world's giant oil fields are given in Table 2.

Presence of an unusual element, mercury, in crude oil from Cymric field of California's San Joaquin Valley has been reported [*2*]. The high level of mercury is attributed to its location on the southeast prolongation of the main mercury belt east of the San Andreas fault. Thus the cinnabar ore found in this region is saturated with hydrocarbons; and the crude oil hydrocarbons are saturated with mercury.

Ratios such as vanadium to vanadium + nickel and iron to vanadium are suggested as being useful for oil type characterizations [*3*]. Eleven North Alaskan oils derived from two genetically distinct source rock sequences were studied for their Cu, Fe, Mn, Ni, and V distribution. Both V-Ni and Ni-API gravity correlations were observed [*3*]. Since organometallic compounds are concentrated in the heavy ends of petroleum, transition element concentrations and ratios can serve as excellent oil–oil correlation parameters.

Generally, vanadium and nickel content increases with asphaltic content of crude oil (API gravity is an indicator). The lighter oils contain less metal [*4*].

The metals present in crudes can be divided into porphyrin and nonporphyrins. The former

[2] This publication, pp. 19-51.

TABLE 1—*Worldwide crude oils* [1].

Country	Characteristics
U.S. (Texas, Louisiana)	Low metal, sulfur, and nitrogen content
U.S. (California)	High metal, sulfur, and nitrogen content
Canada (Alberta)	Similar to U.S. mid-continental
Canada (Western)	Medium-to-high sulfur and trace elements
Venezuela	Medium-to-high sulfur; high in vanadium and trace elements
Colombia	Low in sulfur and trace elements
Africa (Algeria, Libya, Nigeria, etc.)	Low in sulfur and trace elements; high nitrogen
Mid-East (Saudi Arabia, Iran, etc.)	High in sulfur; moderate trace elements
Indonesia	Negligible trace elements

have been the most extensively studied [*4,5*]. Yen [*4*] classifies all metals present in crude oils as follows:

1. **Metalloporphyrin Chelates**
   - Vanadyl and nickel porphyrins
   - Chlorophyll α and other hydroporphins
   - Highly aromatic porphin chelates
   - Porphyrin decomposition ligands
2. **Transition Metal Complexes of Tetradentate Mixed Ligands** (e.g., V, Ni, Fe, Cu, Co, and Cr)
   - Simple complexes from resin molecules
   - Chelates from asphaltene sheets
3. **Organometallic Compounds** (e.g., Hg, Sb, and As)
4. **Carboxylic Acid Salts of Polar Functional Group of Resins** (e.g., Mo and Ge)
5. **Colloidal Minerals** (e.g., silica and NaCl)

TABLE 2—*Elemental analysis of crude oils from giant oil fields* [1].

Country	S, wt%	N, wt%	V, ppm	Ni, ppm	Fe, ppm
Abudhabi	0.62–0.77	0.03	1	0.43	...
Algeria	0.02–0.31	0.01–0.09	0.2–1.5	...	...
Canada	0.2–3.67	0.02–0.12	<1–220	1–60	<1–629
Colombia	0.25–1.11	0.15	59–135	9–14	2–18
Egypt	0.84–2.06	0.08–0.18	15–120	7–72	58
Indonesia	0.07–0.18	0.12–0.34	...	7–33	...
Iran	0.76–3.68	0.02–0.3	11–151	3–39	...
Iraq	1.36–2.1	0.28	6–95	4–15	...
Kuwait	1.82–2.58	0.10–0.13	23–43	6–9	0.3–0.7
Libya	0.13–1.04	0.08–0.27	1–28	1–35	...
Nigeria	0.10–0.26	0.03–0.18	1-4	1-13	...
Saudi Arabia	1.25–3.91	0.03–0.34	6-80	1-20	...
United States					
Alaska (N. Slope)	1.07	0.23	16–30	4–13	...
Arkansas	0.9–2.1	0.02–0.11	12–19	4–23	1–6
California	0.2–1.2	0.1–0.77	1–403	2–174	2–125
Louisiana	0.1–0.8	0.2–0.08	0–4	0–6	...
Oklahoma	0.2–1.18	0.05–0.27	0–148	0–71	1–51
Texas	0.1–2.4	0.01–0.11	0–23	0–5	...
Wyoming	0.1–3.44	0.06–0.36	1–144	1–102	1–23

Using chromatographic separations, gel permeation chromatography, liquid chromatography, and neutron activation analysis, Filby [*6*] has shown that Fe, Co, Cu, Zn, Hg, Cr, and portions of Ni occur as non-porphyrin complexes. Arsenic and Sb appear to be present as low molecular weight compounds such as alkyl or aryl arsines and stibines.

Existence of metal complexes (called porphyrins) in fossil fuels was established by Alfred Treibs in the mid-1930s [*7-10*]. The concept of biomarkers in fossil fuels grew out of Treibs' pioneering work.

Porphyrins are a class of tetrapyrrole macrocyclics with a basic structural unit consisting of four pyrrole rings linked consecutively at the α positions by means of methine bridges (Fig. 1) [*11*]. The principal metal porphyrins in petroleum are Ni + 2 and VO + 2, although

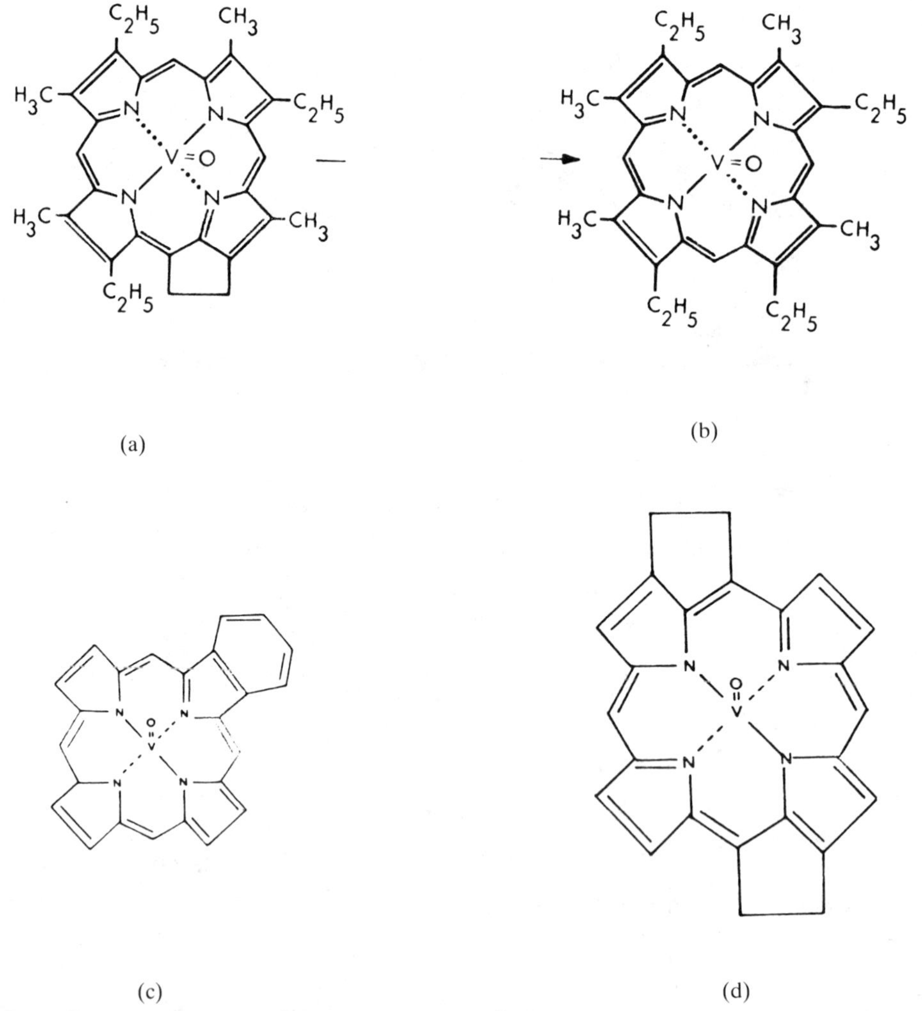

FIG. 1—*Vanadyl porphyrins.* (a) *Deoxophylloerythroetiovanadyl porphyrin.* (b) *Ethiovanadylporphyrin.* (c) *Benzoporphin.* (d) *Porphin structure with two cycloalkano rings.*

Cu + 2 porphyrins may be found in the early stages of sediment diagenesis [*12*]. Major chlorophylls and bacteriochlorophylls appear to be primary source materials of geoporphyrins found in the sedimentary systems. Metal chelation occurs during late diagenesis, predominantly by the fully aromatized porphyrin species. During catagenesis, which is the principal stage of oil generation from kerogen, the distribution and types of metalloporphyrins in the source rock undergo changes due to thermal effects. Metallation results in greater stability of the porphyrin macrocycle and in its preservation throughout the geological record [*12*]. Quirke [*13*] has hypothesized that the presence of metalloporphyrins in sediments requires good ion carriers which can complex with the cyclic tertrapyrroles to form metal complexes with high stability indices and stable metal-nitrogen bonds. This explains the predominance of nickel and vanadyl porphyrins in the geosphere.

A vast body of literature on metal petroporphyrins has been summarized in Refs *4, 5, 14*, and *15*.

An interesting suggestion regarding the presence of up to 0.1 wt% vanadium in some crude oils, particularly Venezuelan Boscan crude, has been to use the resid bottoms as a source of vanadium ore as a byproduct of petroleum refining [*16,17*].

Though tar sands are not strictly a part of crude oils, a brief mention of this topic is appropriate because of its analogous nature. Compared with high sulfur, metal-rich crude oil of the Venezuelan Boscan field, the vanadium content of most tar sand bitumens is relatively low (Table 3) [*18*].

If tar sand bitumens are residues of crude oils that have lost their light ends, it might be expected that metals, which are always concentrated in the heavier fractions of petroleum, would be present in greater amounts than are observed in these bitumens. Metalloporphyrin complexes of vanadium and nickel have been detected in some outcrop and shallow core samples. If these porphyrins are indigenous petroporphyrins, then these compounds can survive the weathering process. Metals in tar sands are always concentrated in the asphaltene fractions. Tar sand extracts high in sulfur usually have V:Ni ratios greater than 1, while those low in sulfur tend to have a V:Ni ratio of less than 118.

Significance of Metals in Petroleum Products

Trace metals in crude oil (Fe, Na, Ni, V, Pb, As) have an adverse affect on the refinery and processing operations, generally acting as catalyst poisons. Vanadium compounds damage the turbine blades and refractory furnaces. Sodium causes problems in furnace brickwork [*20*].

Complexity and expense of the refining operations increases tremendously as the sulfur content of crude oils increase from zero to several wt%. During distillation H2S is evolved and other sulfur compounds concentrate in the heavier distillation residue. The residual fuel

TABLE 3—*Composition of tar sand bitumen.*

	S, %	N, %	V, μmol/g	Ni, μmol/g
Boscan Crude	2.5	0.2	22	20
Tar Sands (Average of 7 from Wo., Utah, Calif., N.M., Alta)	3.0 (0.4–5.7)	0.7 (0.2–1.3)	2.2 (0.5–4.9)	1.3 (0.4–2.5)

oils with more than very small amounts of sulfur have to pay a price penalty in the marketplace. Some local, state, and federal environmental laws ban them altogether [*20*].

Branthaver [*21*] has reviewed the influence of metal complexes, particularly nickel and vanadyl, on various refining operations of crude oils. The effect on processing of heavy crudes is particularly serious.

The sulfur compounds are objectionable in refined solvents since they impart odor and may cause darkening of cooked varnishes and resins [*20*].

The sulfur oxides formed during gasoline combustion can be converted to acids which promote rusting and corrosion of engine parts, piston rings, and cause cylinder wall wear. Sulfur oxides emitted in exhaust are of course atmospheric pollutants [*20*].

In aviation gasoline, sulfur compounds have a deleterious effect on the antiknock efficiency of the alkyl lead compounds. If sulfur content was not limited, specified antiknock values would not be reached for highly leaded grades of aviation fuel. Some sulfur compounds corrode various metal parts of the engine system, with effects varying according to the sulfur compounds present [*20*].

In diesel fuels, vanadium can form low melting compounds (e.g., V205 melts at 691°C) and cause severe corrosive attack on all the high temperature alloys used in the gas turbine blades and diesel engine valves. For example, for AlSl 310 steel, vanadium in fuel should be <2 ppm; at 10 ppm vanadium, the corrosion rate is thrice as high, and at 30 ppm level it is 13 times as much. However, if sufficient magnesium is present in the fuel, it will combine with vanadium to form higher melting compounds and reduce the corrosion [*20*].

Sulfur in diesel fuel causes wear due to the corrosive nature of the combustion byproducts, and increases the amount of deposits in the combustion chamber and on pistons. Low speed diesel engines can tolerate more sulfur than high speed ones [*20*].

High levels of sulfur in fuels adversely affect the carbon forming tendency in gas turbine combustion chambers, and large amounts of sulfur oxides in the combustion gases promote corrosion. Sodium and potassium combine with vanadium to form low melting eutectics (565°C), and with sulfur in the fuel to yield sulfates with melting points in the operating range of gas turbines. They produce severe corrosion. Hence sodium and potassium in the gas turbine fuels must be limited [*20*].

Calcium is not harmful from a corrosion standpoint; actually it helps to inhibit the corrosive action of vanadium. However, calcium can form hard bonded deposits which are not readily removed. Lead can cause corrosion and spoil the beneficial inhibiting effect of magnesium additives on the vanadium corrosion [*20*].

Ash forming metals found in the residual fuel oils are derived normally from the metallic salts found in crude oils. Of these, distillate fuels contain only negligible amounts, with the bulk ending up in the residual fuels. However, both distillate and residual fuels can pick up ash contributors during transportation from the refinery. In water transport in particular, fuel oils can be contaminated with sea water, dirt, scale rust, etc. Ash in heavy fuel oils can cause slagging or deposits and high temperature corrosion in boilers. It may attack the refractories in high temperature furnaces, kilns, etc., and it may affect the finished product in certain industrial processes such as ceramic and glass manufacturing [*20*].

High boiling range fractions and residual fuel usually contain higher amounts of sulfur, which creates corrosion and pollution problems. Conversion of sulfur to SO3 during combustion and later reaction with water will form sulfuric acid, corroding the metal surfaces of equipment. Combining with sodium and vanadium complexes, sulfur forms deposits on the external surfaces of super heater tubes, etc., causing equipment corrosion and loss of thermal efficiency [*20*].

Antiknock additives are added to commercial gasolines to achieve the desired octane levels. The most widely used are tetraethyl and tetramethyl lead. Of course, in recent years, lead is being phased out from gasoline due to environmental concerns and strict government restrictions. Methylcyclopentadienyl manganese tricarbonyl has been studied as one of the octane boosters.

Lubricants and Additives

Unlike the discussion above where the presence of naturally occurring elements in crude oil and refined petroleum causes mechanically adverse effects, a class of organometallic compounds are deliberately added to formulate better lubricants and additives. Generally, the presence of sulfur, chlorine, phosphorus, and alkaline earth metals define the composition of the lubricating oils. Though much of chemistry is proprietary, certain aspects are universal.

Naphthenates and alkyl aryl sulfonates of Ba, Cu, Co, W, Cr, and Fe are added to light fuel oils as combustion catalysts. Naphthenates of Ba, Ca, Mg, or Zn, carbonates of Ca and Mg, and sulfonate or chloride of Mg are added to heavy fuel oils as combustion catalysts, corrosion inhibitors, and antiground additives. Naphthenates or alkylaryl sulfonates of Cu, Fe, or Mn are added as a detergent dispersant component of antiground additives [*11*]. A summary of metal lubricants and additives generally used and their effects are given in Table 4 [*22,23*].

TABLE 4—*Lubricant and additive materials.*

Metal	Compounds	Performance
B	Borax and Esters	Antiwear agents, antioxidant, deodorant cutting oils, greases, brake fluids
Ba	Sulfonates, Phenates, Diorganodiphosphates, Phosphonates, Thiophosphonates	Detergent inhibitors, corrosion inhibitors, detergents, rust inhibitors, ATF, greases
Ca	Sulfonates, Phenates	Detergent inhibitors, dispersants
Cd	Dithiophosphates	Steam turbine oils
Cr	Salts	Grease additive
Hg	Organic Compounds	Bactericide in cutting oil emulsions
Mg	Sulfonates, Phenates	Detergent inhibitors
Mo	MoS_2, Dibutyldithicarbamate, Phosphate	Greases, extreme pressure additives
Ni	Cyclopentadienyl Complexes	Antiwear agents, carbon deposit reduction, improved lubrication and combustion
P	Metaldialkyldithiophosphates	Detergents, antirusting agents
Pb	Naphthenate	Extreme pressure additive, greases, gear oils
Sb	Dialkyldithiocarbamates, Dialkylphosphorodithionates	Antiwear, extreme pressure, antioxidant
Se	Selenides	Oxidation and bearing corrosion inhibitors
Si	Silicone Polymers	Foam inhibitors
Sn	Organo Compounds	Antiscuffing additives, metal deactivators
Zn	Dialkyldithiophosphates, Dithiocarbamates, Phenolates	Antioxidant, corrosion inhibitors, antiwear additives, detergents, extreme pressure additives, crankcase oils, hypoid gear lubricants, greases, aircraft piston engine oils, turbine oils, ATF, railroad diesel engine oils, brake lubricants

Wear Metals in Used Oils

Trace metals in used lubricating oils come from the mechanical wear from oil-wetted components of an engine or as a contaminant from air, fuel, and liquid coolant. Generally, the metals are present as particulate materials rather than as true solutions. The presence of specific metals in the used lubricating oils is associated with specific metal components of an engine. In a normally running engine, wear metal content of the oil slowly increases due to normal wear. However, a sudden increase in one or more metal concentrations in oil indicates failure or excessive wear of a specific engine component. A large body of literature exists defining the analytical determination of wear metals in oils. Table 5 summarizes the typical wear metals and the type of engine wear they indicate [*11,23,24*].

Analysis of Petroleum Products

A vast body of literature abounds concerning the elemental analysis of crude oils, petroleum products, lubricants, and additives. As mentioned earlier, analysis of crude oils and used lubricating oils is an industry-wide standard practice. All the well-established elemental techniques (atomic spectroscopy, ion chromatography, microelemental analysis, neutron activation analysis, spark source mass spectrometry, and X-ray fluorescence) have been utilized for the analysis of petroleum products. It is beyond the scope of this review to exhaustively list all applications. Only typical examples to illustrate the applicability of various techniques will be given.

A number of monographs are available on the subject of elemental analysis of petroleum products [*4,19,25-27*]. Biennial reviews of the journal *Analytical Chemistry* cover the field of petroleum analysis in the odd years and are an excellent source of current literature in this field.

TABLE 5—*Wear metals in used lubricating oils.*

Metal	Wear Indication
Ag	Wrist pin bearings in railroad and auto engines, silver plotted spline lubricating pump
Al	Piston and bearings wear, push rods, air cooler, pump housings, oil pumps, gear castings, box castings
B	Coolant leakage in system
Cd	Bearings
Cr	Ring wear, cooling system leakage, Cr-plated parts in aircraft engines, cylinder liners, seal rings
Cu	Wear in bushings, injector shields, coolant core tubes, thrust washers, valve guides, connecting rods, piston rings, bearings, sleeves, bearing cages
Fe	Wear from engine block, cylinder, gears, cylinder liners, valve guides, wrist pins, rings, camshaft, oil pump, crankshaft, ball and roller bearings, rust
Mg	Cylinder liner, gear box housings in aircraft engines
Mo	Wear in bearing alloys and in oil coolers; various Mo-alloyed components in aircraft engines, piston rings
Na	Antifreeze leakage
Ni	Bearings, valves, gear platings
Pb	Bearings, fuel blowby, thrust bearings, bearing cages, bearing retainers
Sb	Crankshaft and camshaft bearings
Si	Dirt intrusion from improper air cleaner, seal materials
Sn	Bearings and coatings of connecting rods and iron pistons
Ti	Various Ti-alloyed components in aircraft engines
W	Bearings
Zn	Neoprene seals, galvanized piping

However, in spite of the vast analytical technology available and in use for petroleum analysis, ASTM has made only limited use of it in writing new test methods. There are about 40 elemental analysis ASTM methods specifically directed to petroleum related materials (Table 6). Well established techniques, such as inductively coupled plasma emission spectrometry and x-ray fluorescence spectrometry, are only lately being introduced into the petroleum field in ASTM.

In a critical study to assess the capability of analytical techniques and laboratories, U.S. Environmental Protection Agency several years back coordinated the analysis of coal, fly ash, fuel oil, and gasoline samples for 28 elements by 9 reputable laboratories [*28*]. The analytical methods used included neutron activation, atomic absorption, spark source mass spectrometry, optical emission spectrometry, anodic stripping voltametry, and X-ray fluorescence. The results of the study were dismaying, to say the least (Table 7). The wide range in reported concentrations indicates that different sample preparation and analytical techniques can lead to erroneous results.

Sample Preparation

Depending on the technique used for final elemental measurement, the sample preparation can vary from none to quite extensive. For microelemental analysis in general, neutron activation, and some matrices in atomic spectroscopy no sample preparation is necessary. For X-ray fluorescence and heavy matrices in atomic spectroscopy, simple dilution with an organic solvent suffices. For ion chromatography, the sample is combusted in an oxygen environment. For spark source mass spectroscopy and emission spectroscopy, the sample is introduced as a carbon electrode. For certain atomic spectroscopy applications, the sample is wet ashed and brought back into acidic aqueous solution, a time consuming operation.

Dry ashing is generally not preferred since there is a potential loss of elements with high volatility such as As, Cd, Hg, Pb, or Se. Low temperature oxygen plasma ashing, though highly successful for biological samples, has not been very popular in the petroleum analysis area. Volatility losses and ashing time of days are two drawbacks of this system.

Wet ashing utilizing various acids has been employed—sulfuric, nitric, fuming nitric, perchloric, either alone or in mixtures. This digestion tends to minimize volatilization losses [*29*]. Pressurized oxygen bomb decomposition has been utilized to combust the petroleum matrix and release the trace elements [*30*]. However, only less than a gram of sample can be used. The technique can also give rise to metal contamination of the sample from the stainless steel bomb and nickel cups; quartz or Teflon liners and platinum cups have been proposed to reduce the contamination. Several of the refractory and transition metals fail to go into solution after the oxygen bomb combustion. The strength of the bomb technique lies in sample preparation for inorganic anions in petroleum matrix [*31*].

Microwave oven digestions have become popular in recent years [*32*]. However, little work has been done in the petroleum area, primarily because of the safety considerations in exposing such flammable material to microwave radiation, and secondly, due to immiscibility of heavy petroleum materials with acids, only partial recoveries of metals are obtained. It is possible that with the recent availability of microwavable large Teflon closed bombs, petroleum products will be amenable to decomposition by this rapid technique. Several atomic spectroscopic techniques need no sample treatment other than simple dilution with a proper organic solvent before aspiration into a flame or a plasma. The solvents used for this purpose include xylene, kerosene, methyl isobutyl ketone, *n*-heptane, white spirit + isopropanol, toluene, chloroform, *iso*octane, and decalin.

Recently, robots have been utilized for oil dilutions with solvents before introducing into atomic or X-ray spectrometers [*33,34*]. With the aid of a programmed robot, the oil is

TABLE 6—*ASTM test methods for elemental analysis of petroleum products.*

ASTM	IP	Analysis	Technique
D 129	61	Sulfur in petroleum products	Bomb oxidation
D 482	4	Ash from petroleum products	Gravimetry
D 808		Chlorine in new and used petroleum products	Bomb oxidation
D 811		Metals in new and used lube oils	Chemical analysis
D 874	163	Sulfated ash from lube oils and additives	Gravimetry
D 878		Inorganic chloride and sulfate in insulating oils	Gravimetry
D 1018		Hydrogen in petroleum fractions	Combustion-gravimetry
D 1026		Sodium in lube oils and additives	Gravimetry
D 1091		Phosphorus in lube oils and additives	Photometry/gravimetry
D 1266	107	Sulfur in petroleum products	Combustion-gravimetry
D 1317	118	Chlorine in new and used lubricants	Titrimetry
D 1318		Sodium in residual fuel oil	Flame photometry
D 1368		Lead in primary reference fuels	Colorimetry
D 1548		Vanadium in navy special fuel oil	Colorimetry
D 1552		Sulfur in petroleum products	Combustion
D 2547	248	Lead in gasoline	Volumetric
D 2599	228	Lead in gasoline	XRF
D 2622		Sulfur in petroleum products	XRF
D 2785		Trace sulfur	Combustion
D 3116		Lead in gasoline	Colorimetry
D 3120		Sulfur traces in light liquid petroleum	Microcoulometry
D 3227	342	Mercaptan sulfur in petroleum products	Potentiometry
D 3228		Nitrogen in lube and fuel oils	Kjeldahl
D 3229		Lead in gasoline	XRF
D 3231		Phosphorus in gasoline	Colorimetry
D 3237		Lead in gasoline	AAS
D 3340		Lithium and sodium in lubricating greases	Flame photometry
D 3341		Lead in gasoline	Titrimetry
D 3343		Hydrogen in aviation fuels	Correlative
D 3348		Lead in unleaded gasoline	Colorimetry
D 3431		Nitrogen in liquid petroleum hydrocarbons	Microcoulometry
D 3605		Trace metals in gas turbine fuels	AAS/FES
D 3831		Manganese in gasoline	AAS
D 4045		Sulfur in petroleum products	Colorimetry
D 4047	149	Phosphorus in lube oils and additives	Titrimetry
D 4294		Sulfur in petroleum products	XRF
D 4422		Ash in petroleum coke	Gravimetry
D 4628		Ba, Ca, Mg, Zn in unused lube oils	AAS
D 4629	379	Organic nitrogen in liquid petroleum hydrocarbons	Chemiluminescence
D 4927		Additive elements in fresh lube oils	XRF
D 4929		Organic chloride in crude oil	Colorimetry
D 4951		Additive elements in fresh lube oils	ICPAES
Methods under Development:			
		• Nitrogen	Chemiluminescence
		• Chlorine in used oils	Various
		• Carbon/hydrogen/nitrogen	Instrumental
		• Wear metals in used oils	ICPAES

TABLE 7—*Trace element ranges found in EPA round-robin.*[a]

Element	Residual Fuel Oil, ppm	Premium Gasoline, μg/ml
Ag	0.0006–0.1	<0.001–<0.3
As	0.2–<1	<0.02–<200
B	0.002–0.2	<0.02–<0.1
Ba	0.3–5	0.001–<0.02
Be	0.0005–<0.5	<0.0001–<0.01
Ca	7–<400	<0.05–<20
Cd	0.003–1	0.001–<20
Cr	0.7–4	<0.001–<0.3
Cu	0.2–1	0.005–3.8
F	0.004	<0.0003
Fe	10–20	0.1–0.6
Hg	0.002–0.4	0.008–<0.02
K	0.8–5	0.1–<20
Li	0.02–<3	0.001–<0.6
Mg	2–3	0.01–<100
Mn	0.21–1	0.005–<0.1
Na	0.4–30	<0.1–0.6
Ni	20–90	<0.01–<6
P	No analysis	0.2–25
Pb	1–4	2–650
S	No analysis	30–<200
Sb	0.003–<0.5	<0.0004–<1
Se	0.02–0.15	0.001–<0.2
Si	8–30	0.5–<2
Sn	0.01–5	<0.003–<7
Sr	<0.4–<0.5	<0.001–<8
V	40–113	<0.0001–<0.1
Zn	0.4–2	<0.2–3

[a] Excerpted from Ref *28*.

sampled, dispensed into a tared container, diluted with a prefixed amount of solvent, mixed, and then poured into an X-ray cell or aspirated into the flame or plasma nebulizer. Sample throughput and precision of operation are advantages of robotic sample preparations.

Standard Reference Materials

The need for the standardization of analytical methodology and for the wider availability of fossil fuel reference standards was illustrated in the U.S. Environmental Protection Agency's round robin analyses of four fuel matrices for 28 trace elements in 9 laboratories using a variety of analytical techniques [*28*]. Large variations were observed in the reported concentrations of most elements for all samples studied, indicating the errors introduced were due to different sample preparation techniques, interferences peculiar to an analytical technique, and/or operator error. In this area, the role of certified Standard Reference Materials is vital for evaluating the new analytical developments, assessing the lab capabilities, and for routine quality assurance of the data generated in the laboratories.

A number of such standard materials available for the elemental analysis of petroleum products are listed in Table 8. Out of the three sources listed, by far the most stringently examined as well as most widely used standards come from the National Institute of Standards and Technology (formerly National Bureau of Standards). NIST certifies each standard

TABLE 8—*Reference materials for elemental analysis of petroleum products.*

Type	Designation	Certification	Sponsor[a]
• Crude Oil	8505	V	NIST
• Diesel Oil	2821-2829	S	Alpha
	2841-2848	S	Alpha
	#2	S, BTU, CHN	Alpha
	501-458	S	Leco
• Fuel Oil			
–Distillate	1624	S	NIST
–Residual	2811-2817	S	Alpha
	2861-2863	High S	Alpha
	100; 150	S, BTU, CHN	Alpha
	501-095; 502-026	S	Leco
	1619-1623	S	NIST
	1634, A & B	S, Trace Elements	NIST
	1618	V, Ni	NIST
• Kerosene	327	S, BTU, CHN	Alpha
	1616, 1617	S	NIST
• Lube Oil	2831-2837	S	Alpha
	201-203	S, Cl	Alpha
	502-066—071	High S	Leco
	1818	Cl	NIST
	1819	S	NIST
	1083-1085	Wear Metals	NIST
	1836	N	NIST
• Paraffin Oil	501-099	S	Leco
	764-545—547	S	Leco
	501-967—969	S	Leco
• Petroleum Coke	2715-2719; 2720	Ash, S, CHN, Ni, V, Ca, Si	Alpha
	501-479	S	Leco
	501-959—960	S, Ash, CHN, Ni, Fe, V, Ca, Si	Leco
• Reference Fuel	1636-1638; 2712-2715	Pb	NIST

[a] Alpha–Alpha Resources, Stevensville, Mich.
Leco–Leco Corp., St. Joseph, Mich.
NIST–National Institute of Standards and Technology, Gaithersburg, Md.

based on values obtained by at least two independently based methods. It is not clear that the other two commercial sources—Alpha and Leco—follow such painstaking protocol.

In spite of nearly 90 standards listed in Table 8, the majority of them are certified for sulfur. Only a few are available for trace elements. There are none available for additives and lubricating oils. It is hoped that NIST will see this need and fulfill it.

TABLE 9—*Solid fossil fuel SRMs from NIST.*

SRM	NBS Certified + Informational Values	No. of Elements Reported in the Literature
Coal 1632	14 + 7	75
Coal 1632a	19 + 12	68
Coal 1635	14 + 10	59
Fly Ash 1633	12 + 7	73
Fly Ash 1633a	20 + 15	53

At least in the case of NIST SRMs, once they are issued, a body of literature data rapidly builds up increasing the number of elements for which information is available. This has been particularly true in the case of solid fossil fuel SRMs (Table 9) [*35*].

Similarly, the fuel oil 1634 was certified for 7 elements by NBS, but Gladney et al. [*36,37*] list data on 22 elements from literature for this popular SRM.

The SRM 1634 was replaced with 1634a in 1982. However, only 7 trace elements were certified in it. A number of other NIST residual fuel oils (SRM 1619-1623) have been certified for sulfur, but not for any other trace elements. These sulfur standards are potential trace element standards, if homogenity with respect to trace elements can be demonstrated. In one such work, Filby et al. [*38*] using neutron activation analysis have shown that the SRM 1634a is suitable as a standard for 7 more trace elements in addition to those originally certified by NIST. However, 7 other trace elements appeared to be heterogenously distributed and were probably present in mineral particulates. SRM 1620a did not appear to be a suitable standard for any trace element investigated.

Individual analytical methods and their applications to the petroleum related area will be described in Part II of this review.

References

[*1*] Magee, E. M., Hall, H. J., and Varga, G. M., "Potential Pollutants in Fossil Fuels," EPA R2-73-249, National Technical Information Service Publication PB 225-039, 1973.

[*2*] Bailey, E. H., Snavely, P. D., and White, D. E., USGS Paper 424D, U.S. Geological Survey, 1961.

[*3*] Curiale, J. A. in Ref *5*, p. 133.

[*4*] *Role of Trace Metals in Petroleum*, T. F. Yen, Ed., Ann Arbor Science Publishers, Ann Arbor, Mich., 1975.

[*5*] *Metal Complexes in Fossil Fuels*, R. H. Filby and J. F. Branthaver, Eds., ACS Symposium Series No. 344, American Chemical Society, Washington, D.C., 1987.

[*6*] Filby, R. H., Preprint No. 630, Petroleum Division, American Chemical Society, Washington, D.C., 1973.

[*7*] Treibs, A., *Annals of Chemistry*, Vol. 509, 1934, p. 103.

[*8*] Treibs, A., *Annals of Chemistry*, Vol. 510, 1934, p. 42.

[*9*] Treibs, A., *Annals of Chemistry*, Vol. 517, 1935, p. 172.

[*10*] Treibs, A., *Angewandte Chemie*, Vol. 49, 1936, p. 682.

[*11*] Sychra, V., Lang, I., and Sebor, G., *Progress in Analytical Atomic Spectroscopy*, Vol. 4, 1981, p. 341.

[*12*] Filby, R. H. and Van Berkel, G. J. in Ref *5*, p. 1.

[*13*] Quirke, J. M. E. in Ref *5*, p. 74.

[*14*] Yen, T. F. in Ref *4*, p. 1.

[*15*] Filby, R. H. in Ref *4*, p. 31.

[*16*] Fischer, R. P. in Ref *4*, p. 201.

[*17*] Shortsleeve, F. J. in Ref *4*, p. 207.

[*18*] Branthaver, J. F. and Dorrence, S. M. in Ref *19*, p. 141.

[*19*] *Analytical Chemistry of Liquid Fuel Sources*. P. C. Uden, S. Siggia, and H. B. Jensen, Eds., Adv. Chem. Series No. 170, American Chemical Society, Washington, D.C., 1978.

[*20*] *Significance of Tests for Petroleum Products, ASTM STP 7C,* K. Boldt and B. R. Hall, Eds., American Society for Testing and Materials, Philadelphia, 1977.

[*21*] Branthaver, J. F. in Ref *5*, p. 188.

[*22*] Smith, I. C., Ferguson, T. L., and Carson, B. L. in Ref *4*, p. 123.

[*23*] King, A. D., Hilligoss, D. R., and Wallace, G. F., *Atomic Spectroscopy*, Vol. 5, 1984, p. 189.

[*24*] McKenzie, T., Varian Technical Note No. AA10, 1981.

[*25*] Babu, S. P., "Trace Elements in Fuel," American Chemical Society, Washington, D.C., Adv. Chem. Series No. 141, 1975.

[*26*] Hofstader, R. A., Milner, O. I., and Runnels, J. H., "Analysis of Petroleum for Trace Metals," American Chemical Society, Washington, D.C., Adv. Chem. Series No. 156, 1976.

[*27*] Milner, O. I., *Analysis of Petroleum for Trace Elements*, Pergamon Press, Oxford, 1963.

[*28*] Von Lehmden, D. J., Jungers, R. H., and Lee, R. E., *Analytical Chemistry*, Vol. 46, 1974, p. 239.
[*29*] Vigler, M. S., Varnes, A. W., and Strecker, H. A., *American Laboratory*, Vol. 12, No. 8, 1980, p. 31.
[*30*] Parr Instrument Company Bulletin No. 207M.
[*31*] Nadkarni, R. A., *American Laboratory*, Vol. 13, No. 8, 1981, p. 22.
[*32*] Nadkarni, R. A., *Analytical Chemistry*, Vol. 56, 1984, p. 2233.
[*33*] *Advances in Laboratory Robotics*, G. L. Hawk and J. Strimaitis, Eds., Zymark Corp., Hopkinton, Mass., 1984.
[*34*] *Advances in Laboratory Automation Robotics*, G. L. Hawk and J. Strimaitis, Eds., Zymark Corp., Hopkinton, Mass., 1985.
[*35*] Nadkarni, R. A. and Botto, R. I., *Journal of Coal Quality*, Vol. 5, No. 2, 1986, p. 67.
[*36*] Gladney, E. S., LA 8438-MS, 1980.
[*37*] Gladney, E. S., Burns, C. E., Perrin, D. R., Roelandts, I., and Gills, T. E., NBS Special Publication No. 260-88, 1984.
[*38*] Filby, R. H., Van Berkel, G. J., Bragg, A. E., Joubert, A., Robinson, W. E., and Grimm, C. A., *Journal of Radioanalytical and Nuclear Chemistry*, Vol. 91, 1985, p. 361.

R. A. Nadkarni[1]

A Review of Modern Instrumental Methods of Elemental Analysis of Petroleum Related Material: Part II—Analytical Techniques

REFERENCE: Nadkarni, R. A., **"A Review of Modern Instrumental Methods of Elemental Analysis of Petroleum Related Material: Part II—Analytical Techniques,"** *Modern Instrumental Methods of Elemental Analysis of Petroleum Products and Lubricants, ASTM STP 1109,* R. A. Nadkarni, Ed., American Society for Testing and Materials, Philadelphia, 1991, pp. 19–51.

ABSTRACT: A review is presented of the state of the art in elemental analysis of petroleum-related materials (crude oil, gasoline, additives, and lubricants) using modern instrumental analysis techniques. The major instrumental techniques used for elemental analysis of petroleum products include atomic absorption spectrometry (both with flame and with graphite furnace atomizer), inductively coupled plasma atomic emission spectrometry, ion chromatography, microelemental methods, neutron activation, spark source mass spectrometry, and X-ray fluorescence. Each of these techniques is compared for its advantages, disadvantages, and typical applications in the petroleum field.

KEY WORDS: petroleum analysis, atomic spectroscopy, microelemental analysis, X-ray fluorescence, neutron activation analysis, ion chromatography, spark source mass spectrometry

Part I of this paper reviewed the occurrence and significance of trace metals present in petroleum and lubricants.[2] Sample preparation steps were reviewed. In Part II, individual analytical techniques and their applications in the petrochemicals area are discussed.

Atomic Spectroscopic Methods

Without a doubt, the most heavily used elemental analytical technique in the petroleum analysis field is atomic spectroscopy, with flame, nonflame, or plasma atomizers. A number of books [*1*] and reviews [*2–5*] are available. A survey of the vast literature available is beyond the scope of this review; only typical examples to illustrate the applicability of the technique to the field of petroleum analysis will be cited.

Atomic Absorption Spectrometry (AAS)

In the 35 years since Walsh's classic paper [*6*] on AAS this technique has become perhaps the most widely used industrial analytical method. If there is truly a "cook book" type of analysis, it is exemplified by AAS. Many modern AAS instruments are fully automated

[1] Analytical Process Leader, Chemicals Analytical Laboratory, PARAMINS Technology Division, Exxon Chemical Company, Linden, NJ 07036.
[2] This publication, pp. 5–18.

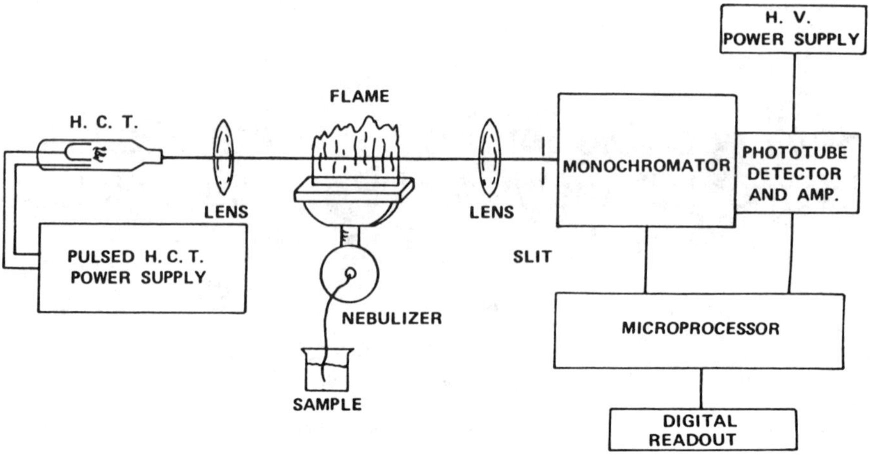

FIG. 1—*Schematic diagram of atomic absorption spectrometer.*

(e.g., determining 6 elements in 50 samples in less than 1 h, with calibration and sample handling). Numerous books, reviews, and thousands of papers are available on AAS.

In AAS, a sample solution is vaporized and the elements atomized at high temperatures (Fig. 1). The element concentration is determined based on the attenuation or absorption by the analyte atoms of a characteristic wavelength emitted from a light source. The light source is typically a hollow cathode lamp which consists of a tungsten anode and a cylindrical cathode enclosed in a gas-tight chamber. The cathode is made up of the analyte metal; usually separate lamps are needed for each element, although some multielement lamps are available. The detector is usually a photomultiplier tube. A monochromator is used to separate the element line, and the light source is modulated to reduce the amount of unwanted radiation reaching the detector. Usually the AAS instruments use flame as the atomization system. The fuel/oxidant composition and ratio determine the temperature of the flame (Table 1).

An air-acetylene flame is used for most elements. The N_2O-acetylene flame is used for more refractory oxide forming elements.

Over 70 elements can be determined by AAS. Nonmetals are usually not determined. Aqueous samples, as well as organic solvents, can be used. The precision of the results is usually on the order of 1 to 5%. Once the sample is in suitable solution, the analysis is completed in a few minutes. AAS offers excellent sensitivity for most elements, with very

TABLE 1—*Flame conditions in AAS.*

Fuel	Oxidant	Temperature, °C	Velocity, cm/s
Natural Gas	Air	1700–1900	55
Propane	Air	1925	82
Hydrogen	Air	2100	320
Acetylene	Air	2300	160
Hydrogen	Oxygen	2550–2700	915
Acetylene	N_2O	2955	180
Acetylene	Oxygen	3050	1130

little spectral interferences. Even though mostly free from atomic spectral interferences, AAS does suffer from molecular spectral interferences, ionization interferences, chemical interferences, and matrix interferences due to differing viscosity and surface tension between sample and the calibrant solutions. The determination of several elements per sample is slow and will require large sample volume due to the sequential nature of the analysis.

Flame Emission Spectrometry (FES) differs from AAS by looking at the emission from an element in the flame rather than its absorption. Thus FES does not need a hollow cathode lamp. FES has superior detection limits to AAS, particularly for alkali and alkaline earth elements. However, for most transition elements, FES has to be carefully used, since it is often difficult to pick out the emission line of interest from nearby potentially interfering lines. Most modern AAS instruments are capable of switching to the FES mode simply by turning a switch.

In Atomic Fluoresence Spectrometry (AFS) the hollow cathode lamp light source is so placed that it does not shine directly into the instrument monochromator. The sample is aspirated into the flame, and the analyte element excited by the light source emits fluorescent radiation. The AFS is claimed to have better detection limits and has potential for multi-element analysis. However, the technique has never reached the commercial success or the wide applicability of atomic spectrometric techniques such as AAS, FES, GFAAS, or-ICPAES.

Applications of AAS in the petroleum field are legion. Books [*1,7*] and excellent reviews [*2–4*] for lube oils and petroleum products exist. Samples that are wet-ashed and brought into aqueous acidic solution can be analyzed by standard AAS methods. Other petroleum samples such as lube oils and additives can be diluted with xylene, MIBK, or kerosene before measurements. In the latter case, organometallic standards are used for calibration. A list of these is given in Table 2, and they are available from the National Bureau of

TABLE 2—*Organometallic AAS/ICPAES standards.*

Element	Compound	Metal, wt%
Ag	Silver 2-ethylhexanoate	42.6
Al	Aluminum cyclohexanebutyrate	5.05
	Aluminum 2-ethylhexanoate	8.07
Ba	Barium cyclohexanebutyrate	28.7
Ca	Calcium 2-ethylhexanoate	12.5
Cd	Cadmium cyclohexanebutyrate	24.8
Co	Cobalt cyclohexanebutyrate	14.8
Cr	Tris(1-phenyl-1,3-butanedione) – Cr (III)	9.6
Cu	Bis(1-phenyl-1,3-butanedione) – Cu (II)	16.37
Hg	Mercuric cyclohexanebutyrate	36.2
Fe	Iron cyclohexanebutyrate	9.9
	Tris(1-phenyl-1,3-butanedione) – Fe (III)	10.45
Mg	Magnesium cyclohexanebutyrate	6.45
Mn	Manganeous cyclohexanebutyrate	13.8
Na	Sodium cyclohexanebutyrate	12.0
Ni	Nickel cyclohexanebutyrate	13.89
P	Triphenylphosphate	9.48
Pb	Lead cyclohexanebutyrate	36.65
Si	Octaphenylcyclotetrasiloxane	14.14
Sn	Dibutyltin bis(2-ethylhexanoate)	22.95
V	Bis(1-phenyl-1,3-butanedione) oxovanadium (IV)	13.01
Zn	Zinc cyclohexanebutyrate	16.66

TABLE 3—*AAS determination of metals in petroleum products.*

Metal	Wavelength, nm	Flame	Detection Limit, μg/mL
Ag	328.1	Air + C_2H_2	0.002
Al	309.3	N_2O + C_2H_2	0.02
As	193.7	Air + C_2H_2	0.2
B	249.8	N_2O + C_2H_2	2
Ba	553.6	N_2O + C_2H_2	0.01
Ca	422.7	Air + C_2H_2	0.002
Cr	357.9	Air + C_2H_2	0.005
Cu	324.8	Air + C_2H_2	0.001
Fe	248.3; 372.0	Air + C_2H_2	0.01
K	766.5	Air + C_2H_2	0.004
Mg	285.2	Air + C_2H_2	0.0001
Mn	279.2	Air + C_2H_2	0.005
Mo	313.3	Air + C_2H_2	0.02
Na	589.0	Air + C_2H_2	0.001
Ni	232.0; 341.5	Air + C_2H_2	0.02
P	213.6	N_2O + C_2H_2	100
Pb	217.0; 283.3	Air + C_2H_2	0.02
Sb	217.6	Air + C_2H_2	0.05
Se	196.1	Air + C_2H_2	0.3
Si	251.6	N_2O + C_2H_2	0.2
Sn	224.6; 286.3; 235.5	N_2O + C_2H_2	0.02
Ti	364.3	N_2O + C_2H_2	0.08
V	318.5	N_2O + C_2H_2	0.08
Zn	213.9; 398.8	Air + C_2H_2	0.003

Standards, Conostan, or Angstrom. Wavelength, flame, and detection limit data for the most commonly determined metals in the petroleum field are given in Table 3.

As to be expected, numerous papers have been published on the determination of lead in gasoline. Gasoline is usually diluted with *iso*octane, MIBK, or acetone [*8*]. Some methods rely on extraction of lead into aqueous solution using, for example, iodine monochloride [*9*] or HCl-HNO_3 mixture [*10*]. In the determination of lead, unlike for other metals, a chemical dependency has been observed. Tetraethyl and tetramethyl lead give different absorptions. This effect can be overcome by the addition of iodine and a quarternary ammonium salt to all samples and standards. Lead determinations have been automated to the extent that over 75 samples can be analyzed each day [*11*].

Viscosity plays an important part in the accuracy of oil analysis. Viscosity differences between sample and calibration standards have been reported to lead to errors of up to 40% for iron [*12*].

Emulsions have also been utilized in petroleum analysis (e.g., lead is determined by using mixtures emulsified by addition of a surfactant and a water-miscible organic solvent such as ethanol or isopropanol [*13*].

Extensive applications of AAS to wear metals analysis is surveyed in a later section.

Graphite Furnace Atomic Absorption Spectrometry (*GFAAS*)

Variously called carbon rod atomization, electrothermal vaporization, or GFAAS, this technique extended the sensitivity of the basic flame AAS technique by several orders of magnitude, and is probably the most sensitive of the atomic spectroscopic family of techniques. Even though capable of analyzing all elements which conventional flame AAS determines, its strength in its applications in the petroleum field lies in the determination of elements which occur at sub ppm levels in petroleum and are difficult to determine by flame AAS.

GFAAS was developed nearly 30 years ago by L'vov [*14*]; however, its widespread use began in the mid-1970s when commercial electrothermal graphite furnace accessories to AAS became available.

The heated graphite furnace consists of a hollow graphite cylinder placed in the light path of an AAS instrument such that the sample light beam passes through it. In the middle of the cylinder top there is a small hole through which up to 100 μL of sample solution is pipetted. The cylinder is purged with nitrogen or argon. An electric current passed through the carbon rod heats the furnace in three stages with increasing temperature: first to evaporate or dry the solvent, second to ash the sample, and last to incandescence to atomize the sample. As the metal is atomized, the transient absorption signal is measured. Temperatures of up to 3000°C are achieved in the final atomization stage.

Like flame AAS, GFAAS is a single element technique. Major advantages of GFAAS are its high sensitivity and small sample volume requirement. Measurement time, however, is longer than that needed for flame AAS. The biggest drawback of GFAAS is that the matrix interference effects are much greater than in flame AAS, which translates into poorer precision (5 to 10% for GFAAS versus 1 to 2% for flame AAS). Because of the sensitivity of the GFAAS technique, extreme care must be taken to observe cleanliness in the laboratory. Dust, water, air, pipette tips, and wipe towels often contribute to variable contamination in the samples. The graphite tubes have a finite lifetime; they usually need to be replaced after about 200 heating cycles.

Lube oil samples can be directly analyzed by GFAAS usually after some dilution with a solvent. Usually 1 to 2 μL of sample is adequate. Since the GFAAS atomization process is independent of the viscosity and surface tension effects that plague pneumatic nebulizers in AAS, petroleum samples can be analyzed using aqueous standard solutions. The organometallic standards used in flame AAS or ICPAES are unnecessary.

A peculiar problem with GFAAS is the low surface tension of the organic solvents used such as xylene or MIBK. The disposable tips of syringes are wetted, and the sample tends to "creep" up the outside of the tips. These type of samples will also wet the inside and ends of the carbon tubes. The problem can be overcome by introducing the sample during the "dry" heating step, so long as the dry temperature is not high enough to melt the plastic tip [*4*].

Analytical conditions for commonly analyzed elements by GFAAS in the petroleum field are summarized in Table 4 [*15*]. Applications of GFAAS in the petroleum field have been reviewed [*2,7,16*]. Even though GFAAS is capable of analyzing nearly 60 elements, the most common applications in the petroleum field are wear metals: V and Ni in crude oils, and Pb, Sb, As, Cd, Cr, Co, Mn, and Mo in petroleum. The 1972 Intercompany Trace Metals project was organized in the United States to develop sensitive analytical methods for low level trace metals in petroleum. This involved five American petroleum companies. Out of this resulted a monograph [*1*] and several papers [*17–19*] that detailed GFAAS methods for Mn, Cd, Sb, Be, Cr, Co, Pb, Mo, Ni, and V. Using $La(NO_3)_3$ as a matrix modifier, phosphorus in gasoline was determined by GFAAS [*20*]. The sensitivity was 20 ng/90 μL sample with a 2% precision. Lead in reference fuel has been determined by converting alkyl lead to inorganic form with bromine solution in CCl_4 and subsequent extraction with 1% HNO_3 for GFAAS measurement. As low as 7 ppb of lead in gasoline was determined [*21*]. Multielement (Fe, Ag, Al, Cr, Cu, Mg, Ni, Si, and Ti) determination

TABLE 4—*GFAAS determination of metals in petroleum products.*[a]

Metal	Wavelength, nm	Matrix Modifier	Ash Temp., °C	Atomzn. Temp., °C	Detection Limits, pg
Ag	328.1	Amm. Phosphate	650	1600	0.5
Al	309.3	$Mg(NO_3)_2$	1700	2400	4
As	193.7	$Ni(NO_3)_2$	1500	2500	20
B	249.7	$Ca(NO_3)_2$	1000	2700	1000
Ba	553.6	...	1200	2700	10
Ca	422.7	...	1100	2600	...
Co	240.7	$Mg(NO_3)_2$	1400	2400	2
Cr	357.9	$Mg(NO_3)_2$	1650	2500	1
Cu	324.7	...	1200	2300	1
Fe	248.3	$Mg(NO_3)_2$	1400	2400	2
K	766.5	...	950	1500	...
Mg	285.2	...	900	2400	0.5
Mn	279.5	$Mg(NO_3)_2$	1400	2200	1
Mo	313.3	...	1800	2700	4
Na	589.6	...	900	1500	...
Ni	232.0	$Mg(NO_3)_2$	1400	2400	10
P	213.6	$La(NO_3)_3$	1400	2700	3000
Pb	283.3	Amm. Phosphate + $Mg(NO_3)_2$	600	1900	5
Sb	217.6	$Ni(NO_3)_2$	1100	2400	15
Se	196.0	$Ni(NO_3)_2$ + $Mg(NO_3)_2$	900	2000	30
Si	251.6	...	1400	2700	40
Sn	224.6; 286.4	Amm. Phosphate + $Mg(NO_3)_2$	1000	2100	20
Ti	365.3	...	1400	2700	50
V	318.4	...	1400	2650	20
Zn	213.9	$Mg(NO_3)_2$	600	1800	0.1

[a] Excerpted from Ref *15*.

of used aircraft lube oils by GFAAS has been achieved with high sample throughput [*22*]. GFAAS is not the ideal technique for the determination of additive elements in lube oils, since the high concentration of metal additives necessitates large dilutions.

Other Nonflame Methods—Only limited applications of hydride generation methods to the petroleum field are available. This includes determination of selenium [*16,23*], lead [*24*], and arsenic [*16*]. Some sample preparation such as oxygen combustion prior to hydride generation is used. Cold vapor atomic absorption for mercury determination in petroleum products has also been described [*15,16*].

Inductively Coupled Plasma Atomic Emission Spectrometry (ICPAES)

The most significant development in the field of atomic spectroscopy in the last two decades has been the development and wide acceptance of the technique of ICPAES. It is second only to GFAAS in its sensitivity to most elements. However, its wide dynamic range and multielement capability makes this technique uniquely suitable for analysis of petrochemicals.

ICPAES was independently developed in two countries: by Greenfield [*26*] in England and by Fassel [*27*] at Iowa State University. A plasma is an electrically neutral, highly ionized gas consisting of ions, electrons, and neutral particles. Argon is the most common gas used. It is ionized by a strong electric field by radio frequency or direct current (DC) plasma, or low power microwave induced plasma (MIP). The DC and microwave plasmas have not reached the popularity of inductively coupled plasma. The RF discharge at 27.12 MHz frequency causes the flowing argon gas to become electrically conductive and reach a temperature up to 15 000°C. The liquid sample sprayed into this plasma through a nebulizer is vaporized, atomized and/or ionized, and electronically excited. The electrons in returning to their ground states emit photons at wavelengths characteristic of an element. The emitted light is measured by a polychromator in a simultaneous spectrometer or by a single photomultiplier in a scanning or sequential spectrometer (Fig. 2).

ICPAES has several advantages over other atomic spectroscopic techniques. It has a linear dynamic range of over five orders of magnitude for many elements. For multielement analysis, this is particularly advantageous, since it permits one to measure percent levels to trace levels of different elements in a sample to be measured simultaneously without dilution. Because of the high temperature of plasma, refractory elements can be easily determined. Alkali metals, however, give poor sensitivity. Again, because of the high temperature, chemical or matrix interferences common in flame AAS are not encountered in ICPAES. For sensitive, rapid multielement analysis, ICPAES is an ideal technique.

Drawbacks of this technique, apart from its high price, are spectral interferences that make accurate trace element determination in complex matrices difficult. High resolution spectrometers are necessary to separate the overlapping spectral lines and correct for background. Sample introduction system continues to be a weakness. Several different types of nebulizers are available, each with its strong points and defects. Aspiration of "high solids" solution is a real problem. The ICPAES detection limits, though good, are not as low as those obtainable by GFAAS.

In the petroleum related field, ICPAES has achieved tremendous popularity, primarily due to its multielement capability and low detection limits. Wavelengths and the dynamic range for the most commonly determined elements in the petroleum area are given in Table 5. Applications of ICPAES in wear metals determination is a particularly productive area and is reviewed in a later section.

As mentioned in an earlier section, the samples may be wet ashed and brought into acidic aqueous solution, or they may be diluted 1:10 with xylene or any suitable solvent before spectrometric measurements. In the latter case, organometallic calibration standards must be used (Table 2). Both methods should yield the same results; for example, in determining nickel and vanadium in NBS residual fuel oil, the results shown in Table 6 were obtained.

Brown [28] determined 21 elements in new and used oils using 1:10 xylene dilution. The solution viscosity was found to cause large errors, errors substantially reduced by adding a

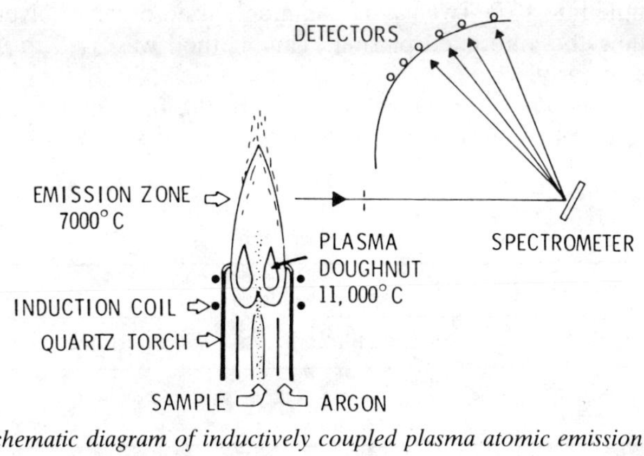

FIG. 2—*Schematic diagram of inductively coupled plasma atomic emission spectrometer.*

TABLE 5—*ICPAES analytics for oil analysis.*

Element	Wavelength, nm	Detection and Upper Limits, ppm
Ag	328.07	0.005–100
Al	308.22; 309.27; 396.15	0.02–100
B	249.77	0.05–100
Ba	455.40, 233.53, 493.41	0.0003–50
Ca	393.37; 317.93; 315.88	0.001–20
Cd	214.44; 226.50; 228.80	0.007–100
Co	228.62	0.007–100
Cr	276.65; 283.56; 267.71	0.01–100
Cu	324.75; 327.40	0.002–50
Fe	259.94	0.004–100
K	766.49	0.2–50
Mg	279.55; 285.21; 279.08	0.0004–20
Mn	257.61; 357.60; 293.30	0.001–50
Mo	202.03; 204.60; 281.62	0.014–100
Na	589.00	0.015–100
Ni	341.47; 231.60; 216.56	0.02–100
P	214.91; 178.29; 177.51	0.10–200
Pb	220.35; 283.31	0.05–100
S	190.00; 180.73; 182.04	0.05–100
Sb	231.15; 206.84	0.05–100
Si	288.16; 251.61	0.02–100
Sn	283.99; 189.99; 242.95	0.05–100
Ti	334.94; 337.28; 350.50	0.001–100
V	292.40; 309.31; 311.07	0.005–100
Zn	206.20; 213.86; 202.55	0.005–20

neutral base oil to the standards. In the additives samples, analyzed by xylene dilution, the presence of detergents in lube oils was found to have only negligible effect [*29*]. Nickel, vanadium, and sulfur in crudes and their products such as residues, bottoms, heavy fuel oils, and gas oils were determined after xylene dilution [*30*]. Klaentschi [*31*] has used low gas flow argon ICP to analyze NBS fuel oils after xylene dilution. Xylene and aqueous solutions could be analyzed at a forward power of 1 kW, but other organic solvents needed 2 kW power; still others have used 4 kW argon-nitrogen ICP to analyze 1:10 diluted xylene solutions [*32*]. Ng and Caruso [*33*] used an electrothermal vaporization cup to introduce 5 μL of organic sample into ICP. Twenty-six commonly used organic solvents were checked without extinguishing the autotuned plasma. This method was used to determine zinc in motor oil and lead in gasoline.

Mackey et al. [*34*] used a Zymark robot to dilute the oil samples with xylene before analysis of additive elements in lubricating oils. With the robot, the results from ICPAES

TABLE 6—*Analysis of residual fuel oil.*

	SRM 1634		SRM 1634A	
	Certified	Xylene Dilution	Certified	Wet Ashing
Ni	36 ± 4	32.7	29 ± 1	28.5 ± 3.7
V	320 ± 15	309	56 ± 2	55.1 ± 4.0

method gave a reproducibility that was at least as good as the routine XRF procedure. For the lighter elements (Mg and P) the repeatability of the ICPAES method was up to five times better.

Usually sulfur is not analyzed by ICPAES unless a vacuum spectrometer is available. Wallace and Ediger describe determination of sulfur in oils using three UV lines [*35*]. Utility of the lines is affected by the plasma background structure. The intensity and complexity of this structure varies with the amount of forward RF power and the solvents used to dilute the oil. Out of 180.73, 182.04, and 182.64 nm sulfur lines, the best line was 180.73 with xylene as solvent at 1250 W of RF power.

Major and trace elements in glycol-based antifreeze fluids have been determined by ICPAES [*36*]. The system had the ability to determine the composition of formulation and information on wear metals simultaneously.

In spite of the ICP multielement capability, some workers have used chemical separations before ICP measurement. Barnes determined eight trace elements in residual fuel oil after separation and preconcentration with a polydithiocarbomate chelating resin [*37*]. The oil was digested with HNO_3 + $HClO_4$ first.

Recently, ASTM Committee D-2 on Petroleum Products and Lubricants completed its first ICPAES standard test method (D 4951) for the determination of additive elements (Ba, Ca, Mg, P, S, and Zn) in lube oils. The samples were diluted with xylene. Results from a round-robin of 17 laboratories analyzing 20 samples are summarized in Table 7. A new round-robin for the determination of wear metals in used oils by ICPAES is in progress.

Wear Metals Analysis

The importance and significance of wear metals analysis of used lube oils has been described in an earlier section. This is a major applications area for atomic spectroscopy in petroleum related materials. There are probably millions of these analyses being done. The U.S. Air Force has conducted this type of analysis on their jet engines for many years under the name "SOAP" (Spectrometric Oil Analysis Program) [*38*]. Excellent reviews of wear metal analysis are available [*2,38*].

Traditionally, wear metals analysis was done by optical emission spectrography, until AAS was introduced to used oil analysis in 1963 [*39*]. In the last decade, ICPAES has become a popular technique for this analysis. Its multielement capability and high plasma temperature make analysis of elements, difficult by AAS, easier.

TABLE 7—*Precision of additive elements analysis of lube oils by WDXRF and ICPAES (1986 ASTM round-robin results).*[a]

Element	Range, wt%	XRF-IS (n = 5)		XRF-MC (n = 7)		ICPAES (n = 12)	
		Repeatability	Reproducibility	Repeatability	Reproducibility	Repeatability	Reproducibility
Ba	0.03–10.0	0.03	0.08	$0.025x^{1.1}$	$0.16x^{0.93}$	0.023x	0.053x
Ca	0.001–1.0	$0.025x^{0.72}$	$0.077x^{0.78}$	$0.028x^{0.75}$	$0.072x^{0.66}$	$0.047x^{0.75}$	$0.12x^{0.75}$
Mg	0.001–1.0	...	...	...	...	$0.022x^{0.67}$	$0.072x^{0.67}$
P	0.01–1.0	$0.0077x^{0.79}$	$0.072x^{0.66}$	$0.044x^{0.92}$	$0.1x^{0.75}$	$0.043x^{0.8}$	$0.11x^{0.8}$
S	0.1–5.0	$0.049x^{0.91}$	$0.2x^{0.94}$	0.04x	$0.16x^{1.04}$	$0.045x^{0.8}$	$0.22x^{0.8}$
Zn	0.001–1.0	$0.012x^{0.73}$	$0.061x^{1.03}$	0.044x	0.11x	0.031x	0.10x

[a] IS = Internal Standard
MC = Mathematical Correction
n = number of participating laboratories

Sampling is particularly a problem with wear metals analysis. Because the sample may contain varying microsize particles, unless aspiration of a timely representative sample takes place, the results will tend to be biased low, since large diameter particles will not be aspirated. Dry as well as wet ashing procedures are devised for this purpose to bring the sample into aqueous solution before spectrometric measurements. Common practice is, however, to dilute the sample directly with an organic solvent (xylene, MIBK, kerosene) before nebulization into a flame or a plasma. Organometallic standards (Table 2) are preferred as the calibration standards.

An extensive review of wear metals analysis, principally by AAS, of pre-1980 papers is given by Sychra et al. [*2*].

A number of papers have compared ICPAES with AAS [*40–42*] or spark rotating disk electrode [*41,42*] methods for wear metals analysis and found them equivalent. A round-robin at 27 laboratories using these three methods found excellent data agreement [*42*]. Generally, xylene, MIBK, or kerosene dilutions have been used for ICPAES [*43,44*]. Smith cautions that wear metal oil samples prepared in xylene should be analyzed within 4 h [*45*]. After 24 h a decrease of ~10% is noted in wear metals concentration. Marked viscosity differences between samples sometimes cause variations in the "green tongue" height of plasma even at 11:1 xylene dilution. In that case a peristaltic pump is used to regulate the sample intake. Some memory effect was observed; a very high level of elements in one sample lingered in the nebulizer for 2 to 3 min. Hence a blank should be run after a "dirty" sample.

Effect of particle size of wear metals continues to be a big concern in atomic spectroscopic analysis. Severely worn engines produce large particulates. It is more important to analyze large particles than small ones in such cases. An appreciable fraction of suspended metallic particles, especially large ones, is not reaching the plasma; rather they collect on the chamber walls and are washed down the drain. Analysis of oils containing particle sizes larger than 3 to 10 μm is not quantitative [*46*]. In one study, a 90 to 100% recovery of particles of <8 μm and a 70 to 100% recovery of 8 μm iron particles were found [*31*].

To overcome this particle size effect, so-called "particle size independent" methods have been developed. These include:

1. Heating oil at 40°C for 5 min with HF + HCl while ultrasonicating and then diluting with kerosene + Neodol 91-6 mixture [*47*].
2. Shaking oil with HF + HNO_3 for 2 min and then diluting with MIBK. As large as 200-mesh molybdenum powder could be quantitatively determined [*48*].
3. Heating oil at 65°C for 45 min with HF + aqua regia while ultrasonicating and then diluting with MIBK and IPA. quantitative recoveries of metal powder suspensions were achieved [*49*].
4. Heating oil at 40°C for 5 min with HF + aqua regia while ultrasonicating and then diluting with MIBK and Neodol 91-6. A number of metals in lube oils and hydraulic fluids were quantitatively determined with 2 to 12% precision [*50*].

Overview of Atomic Spectrometric Methods

Flame AAS, GFAAS, and ICPAES are the most widely used atomic spectrometric methods for petroleum related samples, though AES and DCP methods are still used in specific cases. Table 8 summarizes the detection limits of these more popular techniques. This table is extracted from a review by Parsons et al. [*51*]. As these authors pointed out, several points must be remembered in looking at this table: (1) quantitative determinations cannot be made at the detection limit; (2) relative precision at this level will be ±30 to 50%; and

TABLE 8—*Detection limits of the most popular atomic spectroscopic techniques.*[a]

Element	AAS	AES	GFAAS	DCP	ICP
Ag	0.9	2	0.001	2	0.2
Al	20	3	0.01	2	0.2
As	0.02	2000	0.08	45	2
B	700	50	15	5	0.1
Ba	8	1	0.04	2	0.01
Ca	0.5	0.1	0.01	0.2	0.0001
Cd	0.5	300	0.0002	0.5	0.07
Co	2	5	0.008	1	0.1
Cr	2	1	0.004	1	0.08
Cu	1	3	0.005	2	0.04
Fe	3	10	0.01	3	0.09
K	1	0.01	0.004	0.3	30
Mg	0.1	1	0.0002	0.2	0.003
Mn	0.8	1	0.0005	0.5	0.01
Mo	10	10	0.02	0.5	0.2
Na	0.2	0.01	0.004	0.05	0.1
Ni	2	10	0.05	2	0.2
P	...	100	0.3	75	15
Pb	10	0.2	0.007	23	1
S	...	1600	10	...	30
Sb	0.1	200	0.08	3	10
Se	0.02	...	0.05	45	1
Si	20	...	0.005	15	2
Sn	10	100	0.03	23	3
Ti	10	30	0.3	1	0.03
V	20	7	0.1	8	0.06
Zn	0.8	1000	0.0006	2	0.1

[a] Excerpted from Ref *51*. All values in ng/mL.

(3) for quantitation, 5 to 10 times the level of detection limits are necessary for simple matrices and 20 to 100 times for complex matrices [*51*]. These data along with the detection limits of other popular analytical elemental techniques in the petroleum field are plotted in Fig. 3.

Table 9 summarizes the advantages and drawbacks of the three atomic spectroscopic techniques reviewed above. For lower cost, AAS and GFAAS are good. For best sensitivities, GFAAS and ICPAES are good. For multielement capability and wide dynamic range, ICPAES is unsurpassed.

Spectral interferences in ICPAES, and chemical and matrix interferences in AAS and GFAAS, have to be compensated. A good review of the comparative performances of these methods can be found in an article by Routh [*52*].

ICP–Mass Spectrometry (*MS*)

Less than ten years old, this technique combines the power of ICP for atomizing and ionizing the samples with the sensitivity and selectivity of mass spectrometric detection. Gray demonstrated that useful mass spectra of elemental constituents in solution can be obtained from DC plasma [*53,54*]. The first analytical mass spectra from an ICP was obtained by Houk et al. [*55*]. Sciex, Inc. introduced the first commercial ICP-MS instrument in 1983. Several reviews of this technique have appeared [*56–58*].

ICP is an efficient ion source; however, it operates at atmospheric pressure, while MS needs 10^{-5} torr operating pressure for resolution and detection of ions. This difficulty is

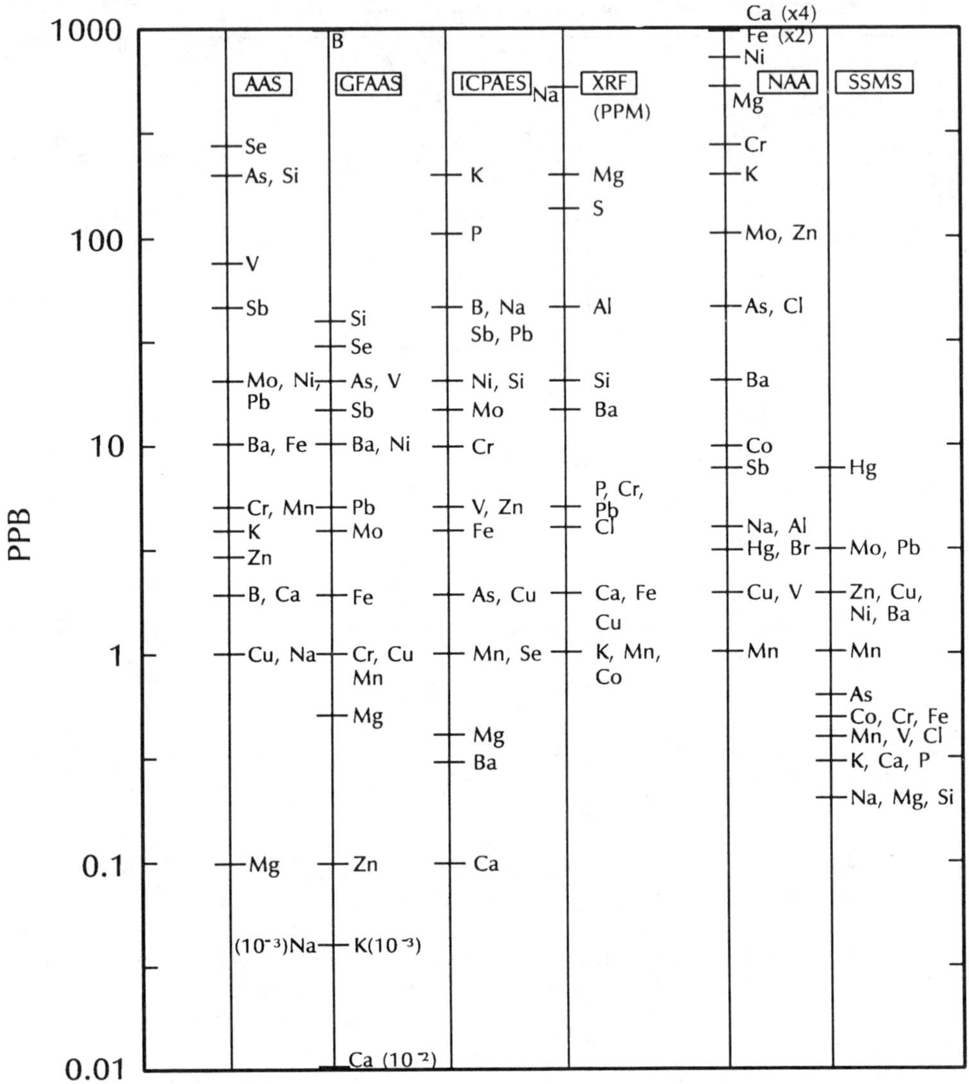

FIG. 3—*Comparison of detection limits for the more common analytical techniques.*

overcome by interfacing the mass spectrometer to the ICP through a series of differentially pumped sampling openings. Ionized atoms from the sample aerosol and other species from plasma enter into a differentially pumped intermediate chamber. There, a small fraction of the analyte ions, together with other background species, enter the low pressure region. The ions of different masses are separated through the quadrupole mass filter. Thus, coupled together, ICP-MS offers an efficient ion source, a less complex background spectrum, the capability of simultaneous determination of over 70 elements in a sample, a wide dynamic range, and detection limits down to tens of parts per trillion in solution for most elements. ICP-MS also offers comparable detection limits for elements, such as arsenic and selenium, normally tested by hydride generation in AAS. In the negative ion mode ICP-MS can

TABLE 9—*Comparison of atomic spectroscopic methods.*

Criteria	AAS	GFAAS	ICPAES
Cost	Moderate	Moderate	High
Maintenance	Moderate	Moderate	High
Dynamic Range	Narrow	Narrow	Wide
Element	Single	Single	Multi
No. of Elements	~70	~70	~70
Detection Limits	ppm	sub-ppb	sub-ppm
Analysis Time	Few minutes	>AAS	Few minutes
Interferences	Chemical, Ionization	Chemical	Spectral
Sample Volume	Few mL	μL	Few mL
Precision, %	1–2	5	5
Applications to Petroleum Area	Wide	A few	Wide

determine sub-ppm levels of halogens [*16*]. Matrix effects are common to ICP-MS just as they are in ICPAES. Solid levels up to 0.1% can be tolerated but, above this level, the analyte ion signal is usually suppressed.

Al-Swaidan [*59*] used dilution of sample in xylene followed by extraction in 40% HNO_3 to determine ppm to sub-ppm levels of Cd, Co, Cu, Fe, Ni, Pb, and Zn in Saudi Arabian crude oil and other gasoline products employing ICP-MS.

Ion Chromatography

Anionic species have been determined for a long time by a variety of wet chemical methods. That situation dramatically changed with the advent of ion chromatography (IC) in 1975 [*60*] and the availability of the first commercial ion chromatograph from Dionex in 1976. Dionex has dominated the field in chemically suppressed IC. Several other companies, such as Wescan, Interaction, Waters, Vydac, Perkin Elmer, and Metrohm, have entered this field with single column nonsuppressed or electronically suppressed IC.

Ion chromatography is a combination of ion exchange chromatography, eluent suppression, and conductimetric detection. For anion analysis, a low capacity anion exchange resin is used in the separator column and a strong cation exchange resin in the H^+ form is used in the suppressor column. A dilute mixture of Na_2Co_3-$NaHCO_3$ is used as the eluent, because carbonate and bicarbonate are conveniently neutralized to low conductivity species and the different combinations of carbonate-bicarbonate give variable buffered pH values. This allows the ions of interest in a large range of affinity to be separated. The anions are eluted through the separating column in the background of carbonate-bicarbonate and conveniently detected based on electrical conductivity. The reactions taking place on these two columns are, for an anion *X*:

Separator Column:

$$\text{Resin} - N^+HCO_3^- + NaX^+ \rightleftharpoons \text{Resin} - N^+X^- + Na^+HCO_3^-$$

Suppressor Column:

$$\text{Resin} - SO_3^- + Na^+HCO_3^- \rightleftharpoons \text{Resin} - SO_3^-Na^+ + H_2CO_3$$

$$\text{Resin} - SO_3^-H^+ + Na^+X^- \rightleftharpoons \text{Resin} - SO_3^-Na^+ + H^+X^-$$

As a result of these reactions in the suppressor column, the sample ions are presented to the conductivity detector as H^+X^-, not in the highly conducting background of carbonate-bicarbonate, but in the low conducting background of H_2CO_3.

Figure 4 shows a schematic representation of the ion chromatography system. A dilute aqueous sample is injected at the head of the separator column. The anion exchange resin selectively causes the various sample anions of different types to migrate through the bed at different respective rates, thus effecting the separation. The effluent from the separator column then passes to the suppressor column where the H^+ form cation exchange resin

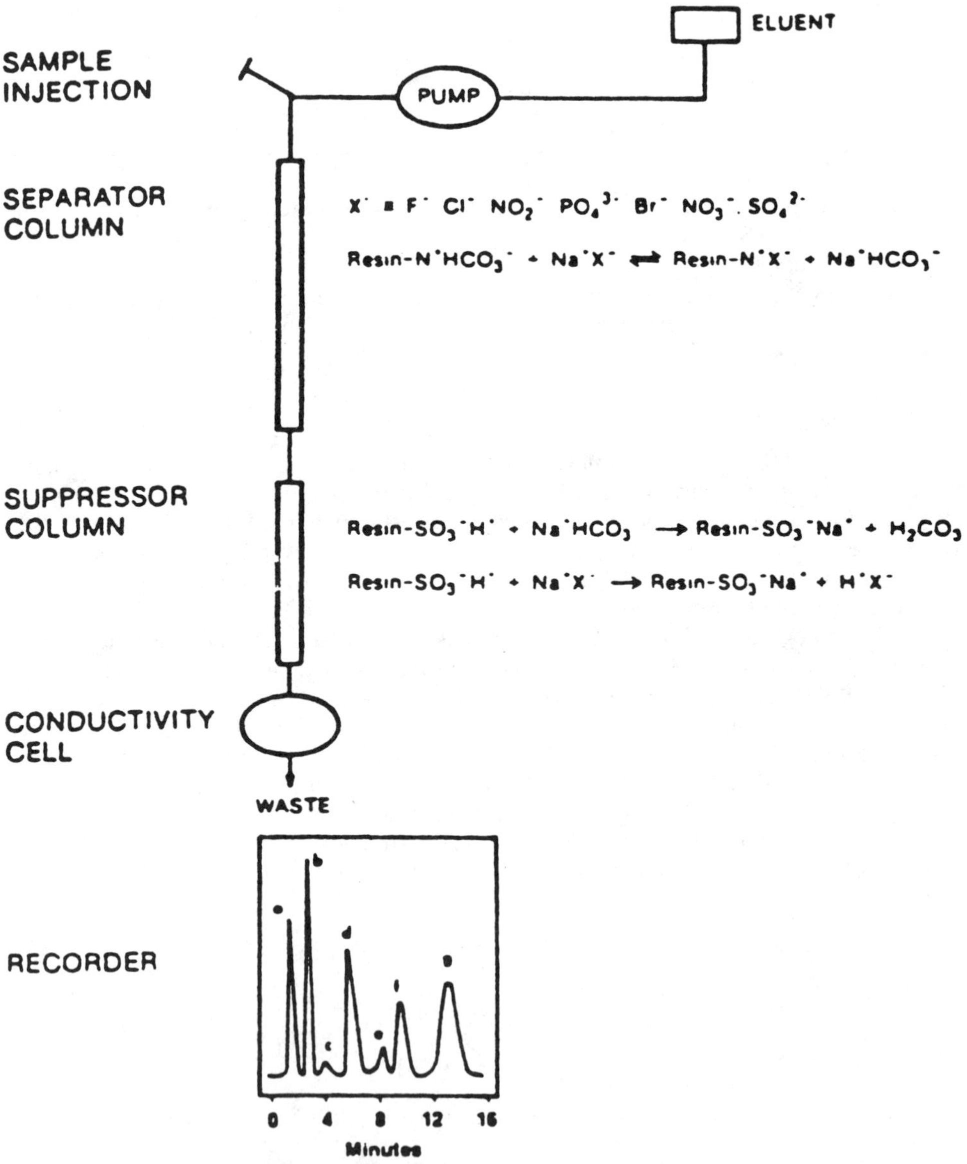

FIG. 4—*Schematic diagram of chemically suppressed ion chromatography.*

absorbs the cations in the eluent stream. Finally, the suppressor column effluent passes through a conductivity cell. The highly conductive anions in a low background conductance of H_2CO_3 are detected at high sensitivity by the conductivity detector. The non-specific nature of the conductimetric detection allows several ions to be sequentially determined in the same sample; at the same time, differential elution rates from the column make the conductimetric detection highly specific and relatively free from interferences. Different stable valence states of the same element can be determined.

On the other hand, because of the nonspecific nature of the conductivity detector, the chromatograph peaks are identified only by their retention times. Thus two ions having the same or close retention times will be detected as one broad peak, giving erroneous results. Figure 5 shows a typical chromatograph for a group of common anions: F^-, Cl^-, NO_2^-, NO_3^-, SO_4^{-2} and $S_2O_3^{-2}$. Table 10 gives retention times and detection limits for the most common anions.

The biggest advantage of IC is its sequential multianion capability. It needs only microlitres of sample. The analysis is rapid (about 10 min for 7 anions). It has a large dynamic range over four decades of concentration. It is also useful for speciation of ions. Some of its disadvantages are: (1) ions with similar retention times interfere with each other; (2) de-

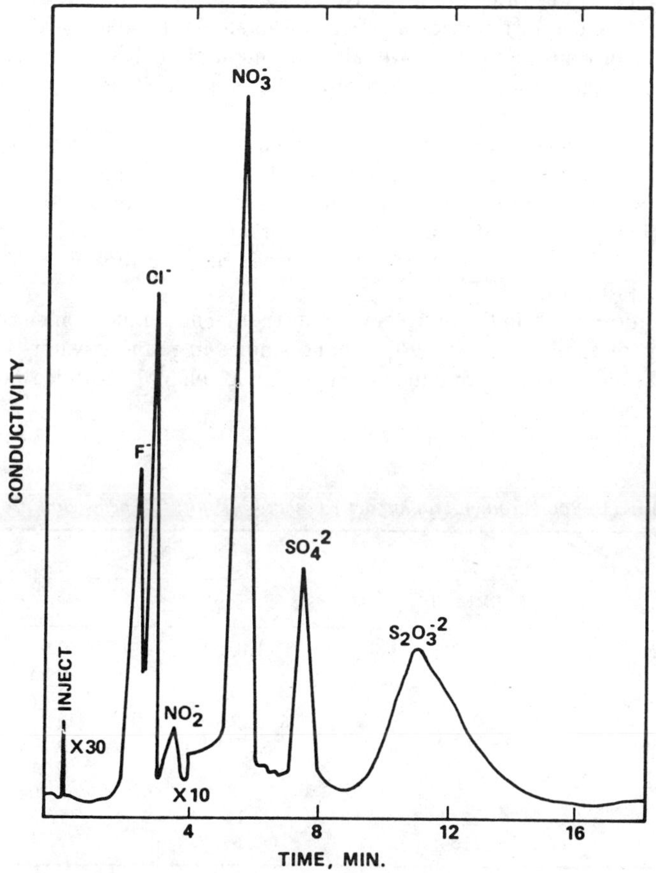

FIG. 5—*Chromatogram of fossil fuel plant sour water.*

TABLE 10—*Ion chromatographic retention times and sensitivity.*[a]

Anion	Retention Time, min	Detection Limit, μg/mL
Fluoride	2.5	0.05
Chloride	3.5	0.07
Nitrite	4.5	0.1
Phosphate	6.0	0.2
Bromide	7.5	0.1
Nitrate	9.0	0.3
Sulfate	10.5	0.1

[a] All measurements are with Dionex IC 16 model; Anion concentrator column, 4 × 50 mm; Anion separator column, 4 × 250 mm; Anion suppressor column, 9 × 100 mm; eluent, 0.0024 *M* Na_2CO_3 + 0.003 *M* $NaHCO_3$; flow rate, 2.3 mL/min; and injection loop, 100 μL.

termination is difficult in the presence of an ion in very large excess; (3) the sample must be an aqueous solution; and (4) the conductimetric detection method is not suitable for anions with pKa of <7. Such ions, however, can be determined using amperometric detection. Even though IC was originally conceived for anionic determinations, cations also can be determined using conductimetric detection for alkali and alkaline earth metals and UV-visible detection for transition metals. Metals determination by IC, however, is not superior to the more traditional methods available for this task. Thus the strong suit of IC is multianion analysis.

A number of IC applications for elemental analysis of petroleum products have been reported. Because the technique is aqueous based, organic samples must be brought into aqueous solution, generally by combustion in oxygen in a Schoniger or Parr combustion bomb. This technique was used by a number of workers to analyze fuel oil, gasoline, diesel fuel, waste oil, coke, and so on for sulfur, fluorine, chlorine, bromine, phosphorus, and nitrogen [*61–69*] (Table 11).

Other applications of IC in the petroleum area (though not in the elemental analysis area) include analysis of engine coolants [*70*], ethanolamines in refinery waters [*71*], fatty acids in drilling mud fluids [*72*], and organic anions, cyanide, phenols, sulfide, ammonia, and so on in refinery wastewaters [*73*].

TABLE 11—*Ion chromatographic determination of sulfur in petroleum products.*

Material	NBS SRM No.	Sulfur, wt%		Ref
		Present	Found	
Residual Fuel Oil	1621	1.05	0.99	*64*
Residual Fuel Oil	1621a	0.94	0.935	*66*
Residual Fuel Oil	1623	0.268	0.271	*61*
		0.268	0.265	*64*
Distillate Fuel Oil	1624	0.211	0.214	*61*
		0.211	0.212	*62*
		0.211	0.204	*66*
Fuel Oil	1634	2.14	2.15	*64*
Fuel Oil	1634a	2.14	2.12	*69*

Microelemental Analysis

This category includes determination of carbon, hydrogen, nitrogen, sulfur, halogens, and oxygen. Usually the sample is combusted or pyrolyzed, the gaseous products formed are absorbed and/or chromatographically separated, and the species of interest are finally quantitated using a variety of detectors such as IR, thermal conductivity, chemiluminescence, and titrimetry.

Carbon-Hydrogen-Nitrogen

There are at least four instruments available for the simultaneous determination of carbon-hydrogen-nitrogen. In the Leco CHN-600 analyzer a sample is burned in tin capsules (1800°C) in pure oxygen, resulting in production of CO_2, H_2O, N_2, N-oxides, and S-oxides. The sulfur oxides are removed with CaO; the excess oxygen and reduction of NO_x to N_2 is achieved with hot Cu; CO_2 is removed with Ascarite (NaOH); water vapor is removed with $MgClO_4$. Finally nitrogen is determined by a thermal conductivity detector, while CO_2 and H_2O are measured with an IR detector (Fig. 6). Up to 200 mg of sample is used. Analysis takes 4 min, and precision better than 2% relative standard deviation can be obtained [*74*].

The Perkin Elmer 2400 analyzer has an operating principle similar to that of Leco CHN-600. The combustion gas stream is cleaned by a successive train of absorbers, and in the end all three components are measured by thermal conductivity detectors.

The Carlo Erba 1106 CHN analyzer is based on the Pregl and Dumas method. The sample is combusted in a tin capsule in pure oxygen (1800°C). Further oxidation is achieved by passing the gaseous mixture over Cr_2O_3 catalyst. From the product gases, hot copper removes excess oxygen and reduces nitrogen oxides to N_2. A chromatographic column of Poropak QS at 100°C separates N_2, CO_2, and H_2O (Fig. 7). The individual gases are measured with a thermal conductivity detector as they are eluted. The method is not very useful for solids, since less than 5 mg of sample is used. Analysis takes about 12 min/sample.

In the Heraeus analyzer, from the combustion gas stream, excess oxygen and NO_x are removed by hot copper, halogens by lead chromate, and sulfur by silver wool. The remaining gases (He, N_2, CO_2, and H_2O) are selectively separated over silica-gel columns at different temperatures. As the absorbed gases are released by changing the temperature, they are measured by a thermal conductivity cell.

Recently ASTM completed a round-robin involving 33 laboratories analyzing 14 petroleum related samples using three of the above analyzers. The samples included crudes, gasolines, fuel oils, naphtha, diesel oil, solvents, lubricants, and additives. The nitrogen method was found to be not applicable to volatile materials and/or those containing <0.1 wt% nitrogen, such as gasoline, jet fuel, naphtha, diesel fuel, or chemical solvents. This problem did not arise for carbon or hydrogen determinations. Table 12 summarizes the data found in this round-robin [*75*].

Nitrogen

In addition to the aforementioned four C-H-N analyzers, there are other instruments available for the determination of nitrogen alone, in the ppm to percent range.

The classic Kjeldahl method of nitrogen determination has been used for over 100 years. A sample is digested with a hot catalyzed mixture of concentrated H_2SO_4 and K_2SO_4 that converts the nitrogen compounds to ammonium salts. These salts are then decomposed in a hot alkaline solution, releasing ammonia, which is then distilled into either standard sulfuric

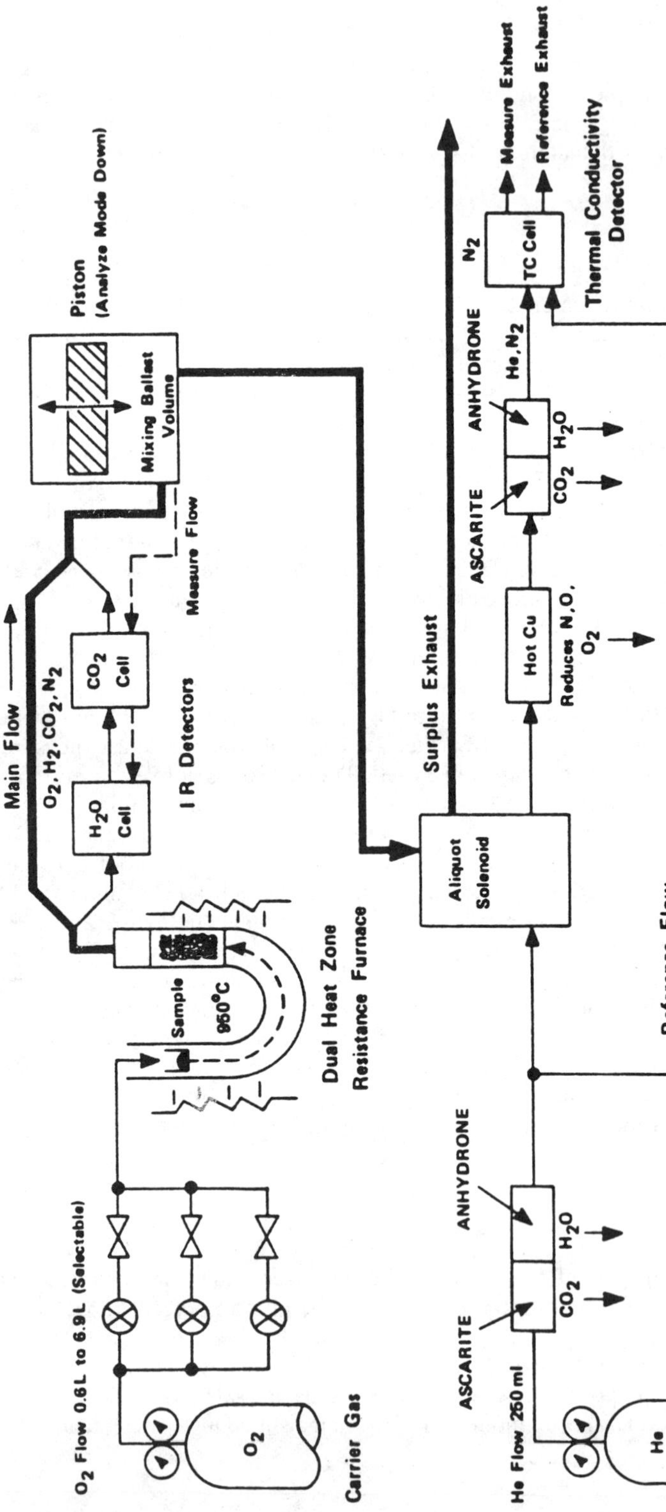

FIG. 6—*Schematic diagram of Leco CHN-600 analyzer.*

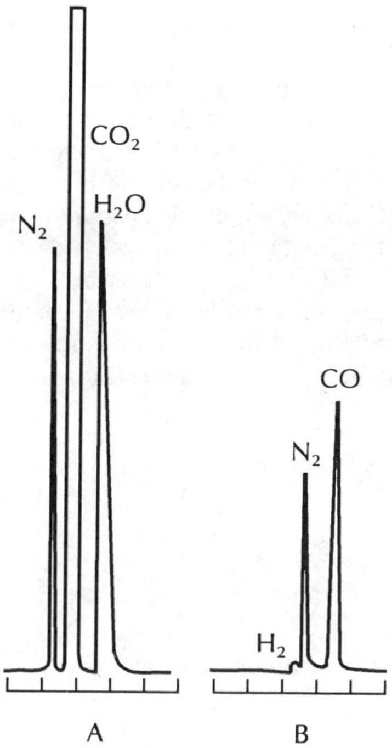

FIG. 7—*Chromatographic separation of combustion gases in a Carlo Erba CHNOS analyzer.* (a) *C-H-N analysis.* (b) *Oxygen analysis.*

or boric acid solution, and finally determined by alkalimetric or acidimetric titration. Mercury, HgO, $HgSO_4$ with selenium, mercuric selenite, and copper selenite have been used as catalysts. Potassium pyrosulfate ($K_2S_2O_7$) helps elevate the boiling point of H_2SO_4 from 340°C to 355–370°C. The high digestion temperature of concentrated H_2SO_4 converts trivalent nitrogen compounds (amines, amides, etc.) as well as most refractory heterocyclic nitrogen compounds (carbazoles, pyridines, aninolines, and pyrolles) to $(NH_4)_2SO_4$. However, the method is not applicable to the compounds containing an N-N linkage, since nitrogen in such compounds may be converted to gaseous nitrogen that will escape detection. The method is also not applicable to oxygen-nitrogen compounds (nitrates) without a special preliminary reduction procedure. Since the modern instrumental C-H-N analyzers combust all forms of nitrogen, they tend to give higher but truer nitrogen values compared to Kjeldahl values. Several automated Kjeldahl analyzers are now commercially available.

In addition to the simultaneous C-H-N analyzers, many vendors also market nitrogen-only analyzers. These are based on similar principles of combustion, chemical absorption,

TABLE 12—*Precision of instrumental carbon-hydrogen-nitrogen analysis.*

Element	Concentration Range, wt%	Repeatability	Reproducibility
Carbon	75–87	$(x + 48.48)0.007151$	$(x + 48.48)0.01805$
Hydrogen	9–16	$(x^{0.5})0.1162$	$(x^{0.5})0.2314$
Nitrogen	0.1–2	0.1246	0.3449

and/or chromatographic separation and thermal conductivity detection. These analyzers can achieve a precision of ±2% RSD and an accuracy from 1 to 5%.

A method particularly useful for trace levels of nitrogen (though it can be used for higher nitrogen levels, by sample dilution) is based on the chemiluminescence principle [*76*]. In these analyzers (Antek and Dohrmann) a sample is introduced into a quartz combustion tube by syringe, by boat inlet, or by a solid sample inlet with a programmable furnace, depending on the sample type. In any event, the sample is heated at 950 to 1100°C in an oxygen-rich atmosphere (Fig. 8). The sample pyrolyzes and combusts with all nitrogen being selectively converted to NO. The gaseous products are conducted through a drier and subsequently reacted in another chamber with ozone to form metastable NO^*_2, which almost instantaneously degrades to its ground state with emission of radiation at 650 to 900 nm. The intensity of this emission (chemiluminescence) is measured with a photomultiplier tube. The reactions are:

$$RN + O_2 \longrightarrow CO_2 + H_2O + NO \text{ at } >950°C$$

$$NO + O_3 \longrightarrow NO^*_2 + O_2$$

$$NO^*_2 \longrightarrow NO_2 + \text{emission at 650 to 900 nm}$$

The technique is capable of determining nitrogen over a wide range of concentrations from ppm to percent. However, the calibration curves or calibration factors are linear over very short ranges, and numerous curves and factors have to be prepared. Alternatively, the sample may be diluted with solvents, so that its diluted concentration falls within some conveniently specified range. The reproducibility of the method is greatly dependent upon the rate at which a liquid sample is injected into the furnace inlet. Generally, multiple injections are made to obtain reproducible and constant results. For liquid injection, a 10 μL sample is used; for syringe boat injection, 10 μL; and for solid boat inlet, a 20 to 30 mg sample is used. The analysis time ranges from <1 min for direct injection to 17 min for solid boat insertion. The precision of either method is ~2% RSD.

Shay and Woodward [*77*] describe the use of an Antek analyzer for petroleum residua (IBP > 300°C) without sample dilution. An accuracy of ±0.3% and a precision of ±1% RSD was achieved for heavy crude oil residues.

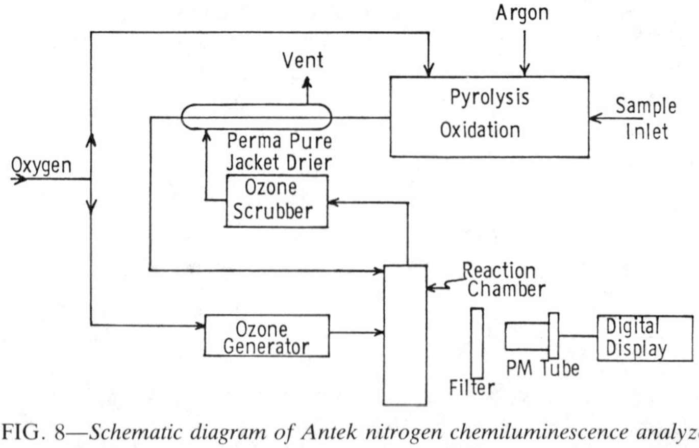

FIG. 8—*Schematic diagram of Antek nitrogen chemiluminescence analyzer.*

Sulfur

Several different types of instrumental analyzers are available for sulfur determination.

Leco Induction Furnace—The sample is burned in a stream of oxygen at a sufficiently high temperature in an induction furnace to convert 95 to 98% of sulfur present to SO_2 (with virtually no SO_3 being produced). SO_2 is then measured by passing the combustion gases into an absorber containing an acid solution of KI and starch indicator. Before the sample is combusted, a small amount of KIO_3 is added to the absorber producing elemental iodine:

$$KIO_3 + 5\ KI + 6\ HCl \longrightarrow 3\ I_2 + 6\ KCl + 3\ H_2O$$

This free iodine produces a blue color with starch. The SO_2 from the combustion gases reacts with free iodine:

$$SO_2 + I_2 + 2\ H_2O \longrightarrow 2\ HI + H_2SO_4$$

and diminishes the intensity of the blue color. A photoelectric cell measuring the intensity of the blue color releases more KIO_3 titrant as the blue color intensity diminishes. When the SO_2 production from the sample ceases and the blue color remains constant, the titration is ended. High SO_2 conversion and minimal SO_3 production is favored at high temperature and low oxygen partial pressure. At 1422°C or higher, the production is 98% SO_2 and is almost independent of oxygen partial pressure. A correction or calibration factor is applied to correct for the SO_3 production. Sometimes the conversion of inorganic sulfates to SO_2 at temperatures <1315°C is incomplete and V_2O_5 needs to be added to aid this conversion. Liquids with boiling points <200°C are not recommended for this analysis. Chlorine <1% does not interfere, but nitrogen >0.1% may, depending upon the types of nitrogen present.

In a modified final measurement step of this procedure, the Leco IR-33 analyzer uses an IR cell for selective SO_2 determination.

Leco SC-32 Analyzer—The sample is combusted in a resistance furnace at 1370°C to convert sulfur to SO_2. Moisture and dust are removed by traps, and SO_2 is measured by an IR detector. About 2 min/sample is needed for analysis. The sensitivity of this instrument is 0.001 wt% sulfur; the precision is about ±2% RSD [*78*].

Carlo Erba 1106 Analyzer—After combustion as in the C-H-N analyzer, the gases are passed over SO_3 catalyst to convert the sulfur compounds to SO_2. Excess oxygen is removed with hot copper, and the gas mixture is separated on a column of Poropak QS. The SO_2 is measured with a thermal conductivity detector. Sample size is <1 mg and a determination takes 5 min. The detection limit is 0.1 wt% sulfur. For ultramicro sulfur analysis, Kirsten and Nordenmark [*79*] modified the procedure so as to use nitrogen as the carrier gas and electron capture detector. The detection limit is 0.0004 wt% in a normal 0.5 mg sample.

Tracor Atlas Sulfur—This analyzer is used for sub-ppm levels of sulfur. Multiple injunctions of 100 μL are used. The light hydrocarbon liquid sample is injected into the hydrogenation tube of a pyrolysis furnace at 1100°C. The hydrogenation products are passed through an acetic acid solution over a lead acetate impregnated sensing tape. The H_2S from the gas stream reacts with lead acetate, darkening the tape; this change is detected by a light beam from a tungsten lamp and is converted into the sulfur concentration. The detection limit of Tracor Atlas is 20 ng sulfur/mL.

Antek Sulfur—This analyzer can be used alone or in conjunction with an Antek nitrogen analyzer and with the same variable types of sample inlet systems. The sample is combusted in a pyroreactor at >1000°C in an oxygen atmosphere. Sulfur containing compounds are converted quantitatively to SO_2. The combustion gases are dried and then flow into a fluorescence chamber where they are exposed to ultraviolet radiation. The fluorescent emission from the following reaction, which is specific for SO_2, is measured with a photomultiplier tube:

$$SO_2 + h\gamma \longrightarrow SO_2 + h\gamma'$$

The linear ranges, detection limit, analysis time, sample size, and precision are similar to those for the Antek nitrogen analyzer.

Halogens

Halogen determination can be done by ion chromatography, neutron activation, or X-ray fluorescence. A specific halogen analyzer can be constructed using a Lindbergh tube furnace. The sample is combusted in oxygen at 1000°C. The HX and X_2 species in gaseous products are converted to halides (X^-) by reduction through an alkaline solution of hydrazine sulfate. The halide is potentiometrically titrated with a standard solution of $AgNO_3$. Less than 0.5 g sample is used. Volatile light liquids and inorganic halides cannot be determined, since they give low recoveries.

Samples combusted in Parr oxygen bomb produce halides as well as other anionic species. From this solution, halides can be determined by ion selective electrodes, ion chromatography, or potentiometric titration with $AgNO_3$.

In commercially available Dohrmann or Mitsubishi microcoulometric analyzers, a sample is injected in a gas stream of O_2 + He in a pyrolysis furnace at ~800°C. The chlorine is converted to chloride and oxychlorides which are reacted with Ag^+ ions in a coulometric cell. The silver ions thus consumed are coulometrically replaced. The total current required to replace the silver ions is a measure of the chlorine present in the injected sample. The detection limit is 200 ppb chlorine.

Oxygen

Some C-H-N instruments such as Carlo Erba or Perkin Elmer can also be used to determine oxygen, based on the Unterzaucher Principle. A sample in a silver capsule is subjected to instantaneous pyrolysis in a helium stream at 1060°C. Nickel-coated carbon catalyst quantitatively converts the gases to CO. This is followed by quantitative chromatographic separation of CO from N_2, CH_4 and H_2 (Fig. 7*b*). CO is measured by a thermal conductivity detector. The analysis takes about 6 min/sample. The results of petroleum products by fast neutron activation and Carlo Erba 1106 analyzer are equivalent (Table 13).

Oxygen by FNAA

Fast Neutron Activation Analysis is widely used for direct oxygen analysis of fossil and synthetic fuels. The analysis of oxygen is based on the nuclear reaction $^{16}O(n,P)^{16}N$. The

TABLE 13—*Determination of oxygen in petroleum products.*

FNAA, %	Carlo Erba, %
0.628	0.630
1.60	1.73
1.18	1.23

fast neutrons are produced by $T(d,n)^4He$ reaction in a Cockcroft-Walton 14 MeV neutron generator. The product ^{16}N has a half-life of 7.2 s and emits 6.1 and 7.1 MeV gamma rays. These high energy gamma rays are far beyond the gamma ray energies of most other elements activated by fast neutrons. The only interferences are:

- $^{16}F(n,\alpha)^{14}N$
- $^{11}B(n,p)^{11}Be$ with a half-life of 13.6 s and a gamma ray of 4.67 MeV

Normally, however, neither of these is a serious interference in petroleum products. The method is a rare example of an analytical technique that is virtually interference free and utilizes a short half-life, thus permitting a very rapid analysis. In practice, about a gram of sample is packed into an oxygen-free small cylindrical polyethylene container (called *rabbit*) and irradiated for 15 s in the neutron beam and then transferred by a fast, pneumatic system to a NaI(Ti) scintillation detector and counted for about 30 s (Fig. 9). The precision of the analysis is dependent on the counting statistics, stability of neutron generation rate, and reproducibility of irradiation delay and counting sequences. The analysis is nondestructive and fast, about 10 min/sample. Repeat analysis on weighed sample requires only a minute.

Neutron Activation Analysis (NAA)

A technique for which a large body of data exists, similar to atomic spectroscopy and X-ray methods in the petroleum field, is NAA. Because of the difficulty of access to a nuclear reactor, the majority of applications of NAA in the petroleum field are of the trace element research type.

Neutron activation analysis, particularly instrumental (INAA), as opposed to NAA with radiochemical separations, has been more extensively used in the last two decades with the wide availability of high resolution Ge(Li) or hyper-purity Ge detectors. In practice about a gram or less of sample sealed in a polyethylene rabbit for short irradiation, or in a quartz capsule for long irradiation, is irradiated along with the elemental standards of interest in a thermal neutron flux of 1×10^{12} n cm^{-2} s^{-1}. Short irradiations last about 1 to 5 min, while long irradiations last up to a week. After each irradiation, the samples are counted on a Ge(Li) detector connected to a multichemical analyzer for varying time periods. Taking

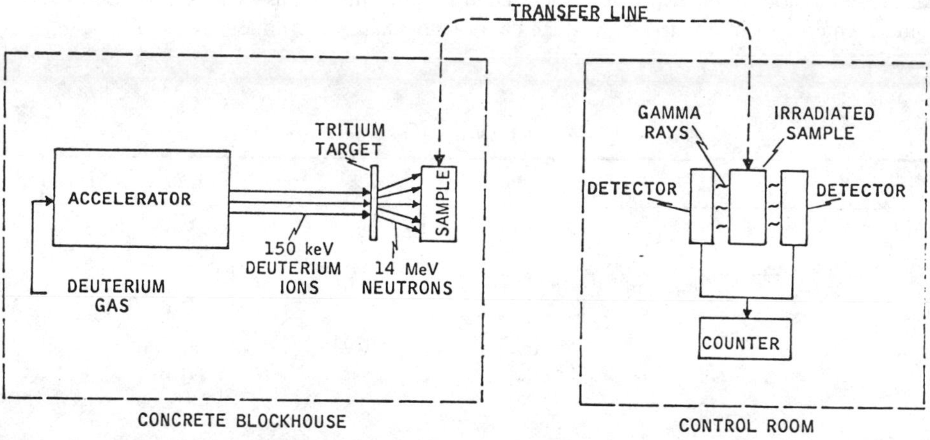

FIG. 9—*Schematic diagram of 14 MeV neutron activation analysis system.*

advantage of different half-lives, up to 40 elements can be measured over a period of 3 to 4 weeks. A scheme for such sequential irradiation and counting is given in Table 14.

Since the petroleum products are essentially a C-H-O matrix that does not get activated during thermal neutron irradiation, there is little background interference. However, some interference can occur from production of the same daughter nuclei from competing *n,p* or *n*,α reactions or from nuclei with the same or similar gamma rays.

Applications of thermal INAA include tar sands [*80–82*] and petroleum and crude oils [*83–94*]. With fast neutrons using a Cockcroft-Walton accelerator, sodium and vanadium in fuel oils have been determined [*85*].

- $^{15}V(n,p)^{51}Ti$; 5.8 min; 0.32 MeV γ
- $^{23}Na(n,\alpha)^{20}F$; 11.6 s; 1.63 MeV γ

Some of the trace element data are used for oil spill identification; see the papers by Filby and Shah [*95*] and Lukens et al. [*96*]. Nearly 300 trace element patterns were accumulated by means of which it can be ascertained whether two samples are from the same or different oil source [*97*].

Only limited use of the ^{252}Cf isotopic source has been made as an activation source in the petroleum field. Meier et al. [*98*] used a 24 μg ^{252}Cf source moderated with paraffin to measure trace amounts of vanadium in oils. A detection limit down to 1 ppm vanadium was achieved. Sulfur in oil has been determined by neutron capture gamma ray spectrometry using a 0.24 μg ^{252}Cf source. This is probably the weakest neutron source used for capture gamma ray analysis. The detection limit was 0.5 wt% sulfur. A stronger source such as 5 μg ^{252}Cf will allow shorter analysis time and improved sensitivity and precision [*99*]. An on-stream vanadium analyzer using ^{252}Cf has been described by Orange and Larson [*100*]. The system design and program for a computer-controlled system for automatic NAA of vanadium in petroleum with a ^{252}Cf source has also been described [*101*]. An unattended system can analyze 48 5-10 g samples in 8 h with a detection limit of 10 ppm ± 30% and accuracies of 0.2 to 4%.

Recent reviews on applications of NAA to fossil fuels include those by Filby [*90*] and Nadkarni [*102*]. In summary, NAA is an extremely sensitive, usually nondestructive technique, generally specific for each analysis, and needing little or no sample pretreatment. On the debit side, it generally needs a nuclear reactor and has associated radiation hazards. Some elements of interest (Pb, S, etc.) cannot be easily determined by NAA.

TABLE 14—*Thermal NAA of petroleum products.*

Ti	Decay Time	Elements Measured
2 min	2–4 min	Al, V, Cu, Ti, S
	10 min	Mg, Ca
	26 min	Cl, I
	30–60 min	Ba, Ni, Mn, K, Cu, Na
8–24 h	2 days	K, Na, Cu
	3–4 days	As, Br
	7 days	Sb, Ca
	15 days	Ba, Cr
	25 days	Fe, Sb, Sr, Ni, Se, Zn, Co

Spark Source Mass Spectrometry (SSMS)

SSMS is one of the most sensitive and comprehensive techniques of trace analysis, particularly for solid samples. Samples are formed into two conducting electrodes held in two movable vises. A radio frequency spark (800 kHz) of high intensity applies a potential, usually in excess of 100 kV, and a discharge is initiated. The repetition rate and duration of the RF spark source is varied to achieve controlled selective volatilization and to avoid overheating the sample. Ion optics are used to collimate the ion beam produced, since the energy spread is broad. Resolution is achieved using double-focusing Mattuch-Herzog geometry (Fig. 10). This allows simultaneous focusing of all masses. Because the ion source is erratic and the ion current fluctuates widely with time, an integrating detector, usually a photographic plate, is employed. After acceleration of the produced ions into the slit system, the electrostatic analyzer selects for transmission ions of a certain energy range with no mass separation. Then the magnetic field separates the ion beam according to the mass:charge ratio. The ions are recorded electronically or photographically. In the latter case the ions cause blackening on a photographic emulsion, which is then developed and evaluated.

Serious interferences in the SSMS spectra occur due to the formation of complex organic masses occurring from 13 to above 200 with considerable overlapping of the trace element lines of interest. These lines are formed from many possible combinations of matrix elements C, H, O, N, and S.

The SSMS technique is capable of the simultaneous determination of over 50 elements from Li to U, in the concentration range from 100 ppm to a few ppb. Typical precisions are ±205% above 1 ppm level and ±10–20% from 1 to 0.001 ppm. Quantitative analysis is carried out by comparing the intensity of the element of interest to that of an internal

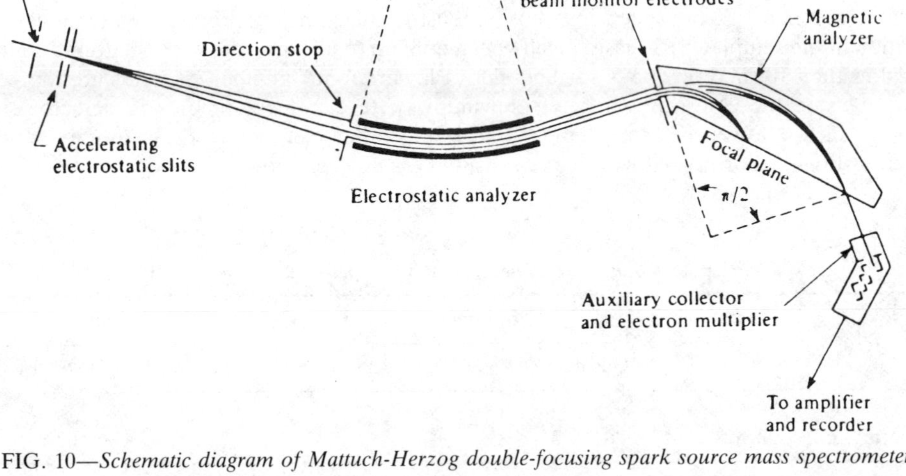

FIG. 10—*Schematic diagram of Mattuch-Herzog double-focusing spark source mass spectrometer.*

matrix element line, usually In or Re. Generally, the elements based on their behavior in SSMS can be divided into three classes:

1. *Analytically Good Elements*—multiisotopic elements such as Li, B, Al, Si, P, Mn, Fe, Cu, Zn, Sr, Mo, Cd, Sn, Ba, and Pb. Precision and accuracy of ~10%.
2. *Analytically Marginal Elements*—monoisotopic elements such as Be, V, Cr, Co, As, Bi, and Sb. If unresolvable interferences occur, concentrations may be only upper limits.
3. *Analytically Poor Elements*—volatile elements such as Na, K, Ca, and Mg. Doubly charged ionic lines of these elements are used. Precision is usually poor. Accuracy varies within a wide range.

Disadvantages of SSMS are time consuming sample preparation, with chance for contamination, extensive molecular and ionic interferences, and poor precision or accuracy (3×) without a standard.

Even though applications of SSMS abound in material sciences, there have been only limited applications in the petroleum field. In recent years, it has been greatly eclipsed by ICPAES or NAA. Lead in gasoline and fuel oil was spiked with ^{204}Pb and treated by either a wet chemical method or bromine oxidation; the separated lead was loaded onto a rhenium filament for mass spectrometric determination [*103*]. The precision of determination is given in Table 15.

X-Ray Fluorescence Spectrometry (XRF)

Perhaps next to atomic spectroscopy, XRF is the most widely used technique for elemental analysis in petroleum products. The technique has the capability of simultaneous multielement nondestructive determination, ppm level sensitivity, and excellent precision and accuracy (~1%). On the other hand, the technique suffers from the high cost of instrumentation (particularly for large multielement capability instruments), various interferences (Compton, absorption, and enhancement), and the problem of matching samples and standards.

The XRF technique is based on the fact that when a sample is bombarded with X-rays, electrons are ejected from the various energy levels of the target atoms. The more energetic electrons in a given atom containing such a vacancy lose energy to fill those shells. In this process X-rays of discrete characteristic energy are emitted. This is known as X-ray fluorescence. Every element has a unique electron distribution; thus every element produces a unique secondary X-ray spectrum whose intensity is proportional to the elemental concentration in the sample. XRF has typically been applied to all elements with an atomic number greater than 10 (neon). The X-ray energies of lower atomic number elements are too weak to penetrate the conventional detector windows. Recent developments in detector technology have allowed measurement of boron and other elements with an atomic number greater than 5 (boron). However, the sensitivity is not very good.

TABLE 15—*Mass spectrometric analysis of petroleum products* [103].

	Pb, ppm		Cd, ppm	
	Gasoline	Fuel Oil	Gasoline	Fuel Oil
SSMS	282 ± 2	0.26 ± 0.03	...	...
Thermal Emission MS	282 ± 2	...	...	0.02 ± 0.002

The basic components of the X-ray spectrometer are a source of excitation consisting of a high energy X-ray tube, an optical system for selecting characteristic X-rays, and integrating or counting circuits for measurements. The X-ray tubes commonly made of tungsten, molybdenum, chromium, rhodium, or scandium have high energy output, spectral purity, and long term stability.

The characteristic lines emitted by the X-ray tube will not only interfere at positions corresponding to their primary wavelengths but at somewhat longer wavelengths due to the Compton effect. Generally, the presence of a characteristic line in the X-ray tube spectrum is expected to cause an error in the determination of a low concentration element whose characteristic line is slightly greater in wavelength than that of the target line.

Due to numerous interelement matrix effects, standards identical in composition to the samples need to be used. Particle size and shape are also important and determine the degree to which the incident beam is absorbed or scattered.

Many of these interferences and matrix effects can be compensated by corrective techniques such as standard addition, internal standards, matrix dilution, thin film method, and mathematical corrections.

Speed, convenience, minimal sample preparation, and its nondestructive nature are the main advantages of the XRF technique. Inherently very precise, XRF rivals the accuracy of wet chemistry techniques in the analysis of major constituents.

WDXRF versus EDXRF

The two main types of XRF instrumentation available can be divided into wavelength (WDXRF) and energy dispersive (EDXRF) instruments. In WDXRF (Fig. 11*a*), the X-rays impinge on the sample and generate fluorescent X-rays. These are then diffracted on a crystal. A goniometer selects the geometry between the crystal and the detector. The geometry controls the detection of X-rays from the element of interest. Different crystals have different sensitivities. Many commercial WDXRF instruments have two detectors and up to six crystals to optimize the instrumental conditions for each element.

The EDXRF instrument uses a much less energetic X-ray tube. The emitted X-ray radiation from the sample impinges directly on a detector, typically a Si(Li), which generates pulses on an incident beam, and these pulses are sorted and counted by a multi-channel analyzer (Fig. 11*b*). Simultaneous determination of all elements (with atomic numbers greater than Mg) is possible. Optimization of specific elements is accomplished through the use of secondary targets and filters. Radioisotope sources can be used in place of X-ray tubes in instruments designed for limited element applications.

Sensitivity in X-ray fluorescence analysis is highly dependent on the sample matrix. Detection limits are typically in the ppm range, with the highest sensitivity observed for the transition metals. Sensitivity declines significantly for the lower atomic number elements. Resolution and sensitivity of EDXRF is typically an order of magnitude worse than for WDXRF. However, the cost of an EDXRF spectrometer, especially the single-element models, can be ten times less than for a WDXRF spectrometer.

Instrumentation

Major vendors of WDXRF instruments are Applied Research Lab, Philips, Rigaku, and Siemens, while main EDXRF vendors are EDAX, EG&G Ortec, Horiba, Kevex, PGT (Outokumpu), and Tracor. Companies that make nondispersive X-ray instruments include Columbia Scientific, Horiba, Oxford, PGT, Texas Nuclear, and Xertec. On-line X-ray instruments are also available from Cianflone and Outokumpu (WDXRF), Texas Nuclear and

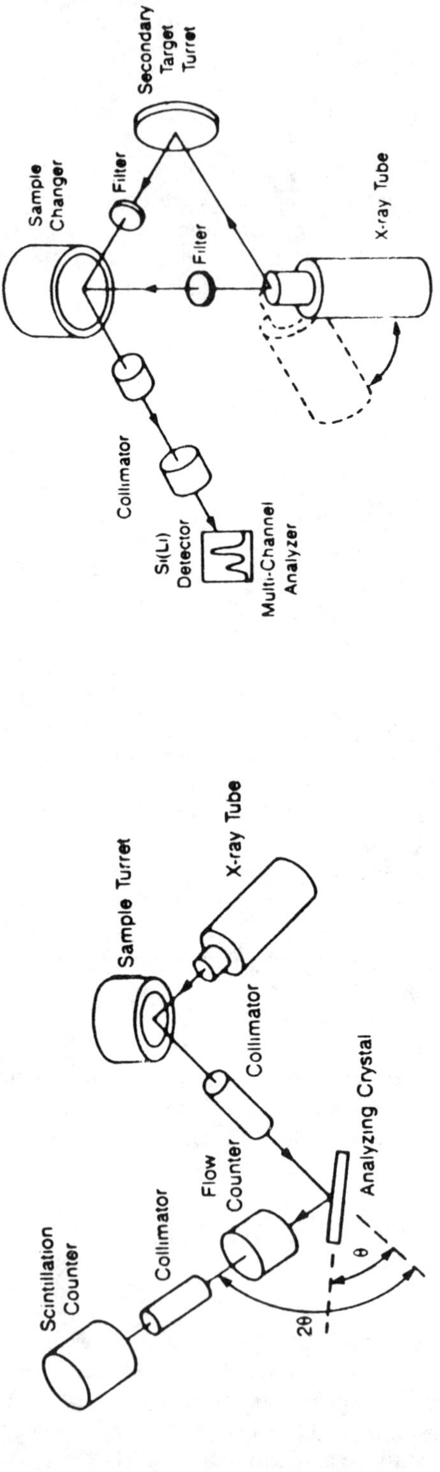

(a) (b)

FIG. 11—*Schematic diagram of X-ray fluorescence analyzer.* (a) *Wavelength-dispersive XRF.* (b) *Energy-dispersive XRF.*

Tracor (EDXRF), and Asoma and PGT (NDXRF). On-line XRF analysis has the advantages of quality control in the plant, elimination of manual laboratory analysis, and possibly better precision of analysis. Applications of on-line X-ray analyzers include lead in gasoline and additives blending of lubricating oils.

Harding et al. [*104*] have described an on-line X-ray analyzer that incorporates a liquid flow cell for the analysis of process streams in the petrochemical industry. Leaded gasoline was analyzed with a detection limit of <10 ppm for elements from V through Zn.

For both WD- and ED- XRF instruments, recent advances include a special window for Si(Li) detectors and the use of layered synthetic microstructures as dispersive elements. These developments should increase the applicability of the XRF technique to lower atomic number elements as well as improve the detection limits.

A newer technique in the XRF area is total reflectance XRF, which does not suffer from the drawback of conventional XRF of incident and fluorescence X-rays strongly interacting with the matrix and giving rise to significant losses. In TRXRF a finely line-focused X-ray beam impinges on a thin film sample at an angle of about 0.05°. This allows total reflection of the incident beam from the sample, resulting in much higher sensitivity, lowered background, and fewer interelement effects.

Interference-free detection limits below 10 pg can be achieved for about 20 elements and below 20 pg for over 40 elements. These correspond to detectable concentrations of 0.1 to 0.2 ppb in aqueous solutions without any preconcentration [*105*]. Potential applications of TRXRF include trace element analysis of environmental, biological, and petroleum samples. Bilbrey et al. [*106*] have determined 10 to 5000 ppm levels of Ti, Cr, Fe, Cu, Pb, V, and Mn in oils using TRXRF. Oils were diluted 10% with chloroform, and 1 μL was applied to the reflector. Relative accuracy was within 10%.

Applications

XRF has been a well established technique in the petroleum industry for a number of years. XRF spectrometers are in routine use in many petrochemical refineries and chemical and blending plants around the world. Some recent applications include: S, Ni, V, Fe, and Cu in crude oils [*107–109*]; Cr, Fe, Ni, Sn, and Pb in aviation oils [*110*]; Pb in fuels [*111*]; V, S, Ni, Mo, Fe, Zn, and Pb in fuel oils [*85,112–114*]; V and S in oils with radioisotope-excited EDXRF [*115*]; and the additive elements (Ba, Ca, Zn, P, Mg, Cu, S, and Cl) in lubricating oils [*34,116,117*]. Both ED- and WD- XRF using fundamental parameters have been employed [*116,117*]. Mackey et al. [*34*] have compared WDXRF with manual and robotic sample preparation ICPAES for the determination of the additive elements Ba, Ca, Cu, Mg, P, and Zn in lubricating oils. XRF was superior to manual, and equivalent to the robotically prepared, ICPAES precision and accuracy.

Wear metals in used lubricating oils also have been determined using XRF [*118–121*]. A 0.45 μm membrane was used to filter and collect all wear metal particles before EDXRF analysis [*119*]. In a different approach the trace metals in oil were deposited onto MgO powder followed by thermal degradation of the organic matrix. The resulting powder was pressed into a pellet and analyzed by EDXRF. Detection limits were a few ppm, and accuracy and precision were on the order of 5 to 10% [*120*].

Recently ASTM Committee D-2 completed a WDXRF standard test method for the determination of additive elements (Ba, Ca, P, S, and Zn) in lube oils (D 4927). Both internal standard and mathematical correction approaches were used. Results from a round-robin of 12 laboratories analyzing 20 oil samples are summarized in Table 7. In general, the repeatability and reproducibility of XRF and ICPAES were on the same order.

As opposed to large XRF spectrometers, several companies (Princeton Gamma Tech,

TABLE 16—*How close do available methods approach the "ideal method" for petroleum analysis?*

Criteria	Ideal	AAS	ICPAES	ICPMS	NAA	SSMS	XRF	IC
Sensitivity	high	good	high	high	high	high	good	good
Dynamic Range	broad	narrow	broad	broad	broad	narrow	broad	broad
Precision	high	high	<AAS	<AAS	high	low	good	good
Interferences	none	some	few	few	some	many	some	some
Sample Preparation	none	some	some	some	none	some	some	>some
Price	low	×	5×	10×	2×	5×	10×	×
Applications	extensive	extensive	extensive	few	large	few	extensive	few
Elements	all	most	most	most	most	all	many	few
Multielements	yes	no	yes	yes	yes	yes	yes	yes
Turnaround	fast	5 min	5 min	5 min	variable	slow	5 min	5 min
Nondestructive	yes	no	no	no	yes	no	yes	no

Horiba, Oxford, Outokumpu, and Xertex) have made available single element or a few element capable small X-ray analyzers. They are considerably cheaper than the large analyzers and have found a niche in small laboratories for analysis of a few specific elements in petroleum related samples. Most of these analyzers are desk-top size. Precisions of 1% can be obtained. Usually the elements determined are chlorine, sulfur, lead, and additive elements in lubricating and cutting oils. Most of the publications in this area are from vendor companies. Examples include determination of Zn, Ca, S, and P in lube oils [*122*] and oil additives [*123*] with an Oxford Lab X2002 analyzer using Cd^{109} excitation source for zinc and Fe^{55} for other elements; determination of S, Ca, Zn, and Ba in lubricating oils [*124*] and S and Cl in cutting oils [*125*] with a Princeton Gamma Tech/Outokumpu X-Met 820 portable analyzer using Fe^{55} as the excitation source for light elements such as S, Ca, Ba, and Cl and Cm^{244} for heavy elements such as zinc; and determination of sulfur in various oils with a Xertex 702 analyzer using a titanium target X-ray tube operated at 7.5 kV and 0.4 MA with a neon-filled proportional counter detector [*126*].

Concluding Remarks

A number of analytical techniques are widely used for the elemental analysis of petroleum related materials. Most of these techniques are fully matured and have been extensively documented in the literature. Yet no one technique is ideal for all types of analysis. An analyst has to make a pragmatic choice based on specific criteria for a specific matrix for a particular analysis. Hieftje [*127*] has listed a number of criteria for choosing an ideal analytical technique: high sensitivity, broad linear dynamic range, high precision, no matrix interferences, inexpensive to acquire and to operate, useful to all elements, simple to operate, compact, minimum sample preparation, microsampling capability, rapid, nondestructive, etc. Table 16 summarizes how individual techniques fare against an "ideal" technique.

Few if any of the methods discussed in this review can meet all these criteria. However, we are fortunate in the petrochemicals analyses field that many of the methods we use meet several of these criteria, and an analyst can wisely make a choice as to the best available technique for a particular analysis.

Acknowledgments

The wholehearted support of my colleagues Carol Murdichian, Catherine Schenck, and Ralph Spohn in preparation of this review is greatly appreciated.

References

[1] Hofstader, R. A., Milner, O. I., and Runnels, J. H., "Analysis of Petroleum for Trace Metals," American Chemical Society, Washington, D.C., Adv. Chem. Series No. 156, 1976.
[2] Sychra, V., Lang, I., and Sebor, G., *Progress in Analytical Atomic Spectroscopy,* Vol. 4, 1981, p. 341.
[3] de al Guardia, M. and Salvador, A., *Atomic Spectroscopy,* Vol. 5, 1984, p. 150.
[4] Pradhan, N. K., *Coal and Petroleum Analysis by Atomic Absorption Spectroscopy,* Varian Techtron, Springvale, 1976.
[5] Varnes, A. W., *Spectroscopy,* Vol. 1, No. 11, 1986, p. 28.
[6] Walsh, A., *Spectrochimica Acta,* Vol. 7, 1955, p. 108.
[7] Van Loon, J. C., *Analytical Atomic Absorption Spectroscopy: Selected Methods,* Academic Press, New York, 1980.
[8] McCorriston, L. L. and Ritchie, R. K., *Analytical Chemistry,* Vol. 47, 1975, p. 1137.
[9] Coleman, M. F. M., *Journal of Chemical Education,* Vol. 62, 1985, p. 261.
[10] Bowen, B. C. and Foote, H., Institute of Petroleum Report IP 74-010, London.
[11] Heistand, R. N. and Shaner, W. C., *Atomic Absorption Newsletter,* Vol. 13, 1974, p. 65.
[12] Ibrahim, R. J. and Sabbah, S., *Atomic Absorption Newsletter,* Vol. 14, 1975, p. 131.
[13] Cardarelli, E., Cifani, M., Mecozzi, M., and Schi, G., *Talanta,* Vol. 33, 1986, p. 279.
[14] L'Vov, B. V., *Spectrochimica Acta,* Vol. 17, 1961, p. 761.
[15] Slavin, W., "Graphite Furnace AAS: A Source Book," Perkin Elmer Corp., Norfolk, Conn., 1984.
[16] Meyer, G. A., *Analytical Chemistry,* Vol. 59, 1987, p. 1345A.
[17] Robbins, W. K., *Analytical Chemistry,* Vol. 46, 1974, p. 2177.
[18] Runnels, J. H., Merryfield, R., and Fisher, H. B., *Analytical Chemistry,* Vol. 47, 1975, p. 1258.
[19] Robbins, W. K. and Walker, H. H., *Analytical Chemistry,* Vol. 47, 1975, p. 1269.
[20] Driscoll, D. J., Clay, D. A., Rogers, C. H., Jungers, R. H., and Butler, F. E., *Analytical Chemistry,* Vol. 50, 1978, p. 767.
[21] Epstein, M. S., *Atomic Spectroscopy,* Vol. 4, 1983, p. 62.
[22] Niu, W., Haring, R., and Newman, R., *American Laboratory,* Vol. 19, No. 11, 1987, p. 40.
[23] Walker, H. H., Runnels, J. H., and Merryfield, R., *Analytical Chemistry,* Vol. 48, 1976, p. 2056.
[24] Aznarez, J., Vidal, J. C., and Carciner, R., *Journal of Analytical Atomic Spectrometry,* Vol. 2, 1987, p. 55.
[25] Knauer, H. E. and Milliman, G. E., *Analytical Chemistry,* Vol. 47, 1975, p. 1263.
[26] Greenfield, S., Jones, I. L., and Berry, C. T., *Analyst,* Vol. 89, 1964, p. 713.
[27] Wendt, R. H. and Fassel, V. A., *Analytical Chemistry,* Vol. 37, 1965, p. 920.
[28] Brown, R. J., *Spectrochimica Acta,* Vol. 38B, 1983, p. 283.
[29] Merryfield, R. N. and Loyd, R. C., *Analytical Chemistry,* Vol. 51, 1979, p. 1965.
[30] Fabec, J. L. and Ruschak, M. L., *Analytical Chemistry,* Vol. 57, 1985, p. 1853.
[31] Klentschi, N., *Materials Technology,* Vol. 12, 1984, p. 3.
[32] Broekaert, J. A. C., Keis, F., and Laqua, K., *Talanta,* Vol. 28, 1981, p. 745.
[33] Ng, K. C. and Caruso, J. A., *Analytical Chemistry,* Vol. 55, 1983, p. 2032.
[34] Mackey, J. R., Meunier, C. A., and Windsor, B. K., paper presented to Winter Conference on Plasma Spectrochemistry, San Diego, Calif., Jan. 1988.
[35] Wallace, G. F. and Ediger, R. D., *Atomic Spectroscopy,* Vol. 2, 1981, p. 169.
[36] Beckwith, P. M., *Spectrochimica Acta,* Vol. 40B, 1985, p. 301.
[37] Mahanti, M. S. and Barnes, R. M., *Analytical Chimica Acta,* Vol. 149, 1983, p. 395.
[38] Eisentraut, K. J., Newman, R. W., Saba, C. S., Kaufman, R. E., and Rhine, W. E., *Analytical Chemistry,* Vol. 56, 1984, p. 1086A.
[39] Sprague, S. and Slavin, W., *Atomic Absorption Newsletter,* Vol. 2, 1963, p. 20.
[40] McKenzie, T., Varian Technical Note No. AA-IO, 1981.
[41] Varnes, A. W. and Andrews, T. E., *Jarrell-Ash Plasma Newsletter,* Vol. 1, No. 1, 1978, p. 12.
[42] Allied Analytical Systems ICP Application Note, 1985.
[43] Fassel, V. A., Peterson, C. A., Abercrombie, F. N., and Kniseley, R. N., *Analytical Chemistry,* Vol. 48, 1976, p. 516.
[44] Ward, A. F. and Marciello, L., *Jarrell-Ash Plasma Newsletter,* Vol. 1, No. 4, 1978, p. 10.
[45] Smith, J. C., *Jarrell-Ash Plasma Newsletter,* Vol. 2, No. 3, 1979, p. 4.
[46] Saba, C. S., Rhine, W. E., and Eisentraut, K. J., *Analytical Chemistry,* Vol. 53, 1981, p. 1099.
[47] Nygaard, D. D., Chase, D. S., and Leighty, D. A., Instrumentation Lab Application Note No. AID 183.
[48] Saba, C. S. and Eisentraut, K. J., *Analytical Chemistry,* Vol. 51, 1979, p. 1927.

[49] Brown, J. R., Saba, C. S., Rhine, W. E., and Eisentraut, K. J., *Analytical Chemistry,* Vol. 52, 1980, p. 2365.
[50] Kauffman, R. E., Saba, C. S., Rhine, W. E., and Eisentraut, K. J., *Analytical Chemistry,* Vol. 54, 1982, p. 975.
[51] Parsons, M. L., Majors, S., and Forster, A. R., *Applied Spectroscopy,* Vol. 5, 1983, p. 411.
[52] Routh, M. W., *Spectroscopy,* Vol. 2, No. 2, 1987, p. 45.
[53] Gray, A. L., *Analyst,* Vol. 100, 1975, p. 289.
[54] Gray, A. L., *Analytical Chemistry,* Vol. 47, 1975, p. 600.
[55] Houk, R. S., Fassel, V. A., Flesch, G. D., Svec, H. J., Gray, A. L., and Taylor, C. E., *Analytical Chemistry,* Vol. 52, 1980, p. 2283.
[56] Date, A. R. and Gray, A. L., *Spectrochimica Acta,* Vol. 38B, 1983, p. 29.
[57] Taylor, H. E., *Spectroscopy,* Vol. 1, No. 11, 1986, p. 20.
[58] Houk, R. S., *Analytical Chemistry,* Vol. 58, 1986, p. 97A.
[59] Al-Swaidan, H. M., *Analytical Letters,* Vol. 21, 1988, p. 1487.
[60] Small, H., Stevens, T. S., and Bauman, W. C., *Analytical Chemistry,* Vol. 47, 1975, p. 1801.
[61] Mizisin, C. S., Kuivinen, D. E., and Otterson, D. A. in *Ion Chromatographic Analysis of Environmental Pollutants,* J. D. Mulik and E. Sawicki, Eds., Ann Arbor Science, Ann Arbor, Mich., Vol. 2, 1979, p. 129.
[62] Butler, F. E., Toth, F. J., Driscoll, D. J., Hein, J. N., and Jungers, R. H. in *Ion Chromatographic Analysis of Environmental Pollutants,* J. D. Mulik and E. Sawicki, Eds., Ann Arbor Science Publishers, Ann Arbor, Mich., Vol. 2, 1979, p. 185.
[63] Koch, W. F., *Journal of Research of NBS,* Vol. 84, 1979, p. 241.
[64] McCormick, M. J., *Analytical Chimica Acta,* Vol. 121, 1980, p. 233.
[65] Saitoh, H. and Oikawa, K., *Bunseki Kagaku,* Vol. 31, 1982, p. E375.
[66] Viswanadham, P., Smick, D. R., Pisney, J. J., and Dilworth, W. F., *Analytical Chemistry,* Vol. 54, 1982, p. 2431.
[67] Dionex Application Note No. 15, 1979.
[68] Wetzel, R., Smith, F. C., and Cathers, E., *Industrial Research and Development News,* Vol. 23, No. 1, 1981, p. 152.
[69] Nadkarni, R. A. and Pond, D. M., *Analytica Chimica Acta,* Vol. 146, 1983, p. 261.
[70] Dionex Application Note No. 4, 1978.
[71] Dionex Application Note No. 39, 1982.
[72] Dionex Application Note No. 45, 1983.
[73] Nadkarni, R. A. and Brewer, J. M., *American Laboratory,* Vol. 19, No. 2, 1987, p. 50.
[74] Nadkarni, R. A., Cornett, R. B., and Bredweg, R. L., *American Laboratory,* Vol. 17, No. 2, 1985, p. 69.
[75] Nadkarni, R. A. and Bover, W. J., ASTM Research Report, forthcoming.
[76] Druschel, H. V., *Analytical Chemistry,* Vol. 49, 1977, p. 932.
[77] Shay, J. Y. and Woodward, P. W., *American Laboratory,* Vol. 18, No. 10, 1986, p. 114.
[78] Nadkarni, R. A., Hebel, M. T., and Taunton, J. W., *Journal of Coal Quality,* Vol. 4, No. 1, 1985, p. 72.
[79] Kirsten, W. J. and Nordenmark, B. S., *Analytica Chimica Acta,* Vol. 196, 1987, p. 59.
[80] Filby, R. H. and Jacobs, F. S., *Analytica Chemica,* Vol. 55, 1983, p. 74.
[81] Jervis, R. E., Ho, K. L. R., Tiefenbach, B., *Journal of Radioanalytical Chemistry,* Vol. 71, 1982, p. 225.
[82] Jacobs, F. S. and Filby, R. H. in Ref *90,* p. 49.
[83] Shah, K. R., Filby, R. H., and Haller, W. A., *Journal of Radioanalytical Chemistry,* Vol. 6, 1970, p. 185.
[84] Ibid., p. 413.
[85] Persiani, C. and Shelby, W. D., *Environmental Science and Technology,* Vol. 7, 1973, p. 125.
[86] Colombo, U. P., Sironi, G., Fasolo, G. B., and Malvaro, R., *Analytical Chemistry,* Vol. 36, 1964, p. 802.
[87] Yule, H. P., *Nuclear Applications,* Vol. 3, 1967, p. 637.
[88] Flaherty, J. P. and Eldridge, H. B., *Applied Spectrometry,* Vol. 24, 1970, p. 534.
[89] Al-Shahristani, H. and Al-Atyia, M. J., *Journal of Radioanalytical Chemistry,* Vol. 14, 1973, p. 401.
[90] Filby, R. H. et al., Eds., *Atomic and Nuclear Methods in Fossil Energy Research,* Plenum Press, New York, 1982.
[91] Buenafama, H. D. in Ref *90,* p. 29.
[92] Filby, R. H. and Shah, K. R. in *Role of Trace Elements in Petroleum,* T. F. Yen, Ed., Ann Arbor Science Publishers, Ann Arbor, Mich., 1975, p. 89.

[*93*] Ndiokwere, C. L., *Radiochemical and Radioanalytical Letters,* Vol. 59, 1983, p. 201.
[*94*] Block, C. and Dams, R., *Journal of Radioanalytical Chemistry,* Vol. 46, 1978, p. 137.
[*95*] Filby, R. H. and Shah, K. R. in *Nuclear Methods in Environmental Research,* Vol. 1, 1971, p. 86.
[*96*] Lukens, H. R., Bryan, D., Hiatt, N. A., and Schlesinger, H. L., Gulf-RT-A-10684, 1971.
[*97*] Lukens, H. R., Gulf-RT-A-10973, 1972.
[*98*] Meier, H., Zeitler, G., and Menge, P., *Journal of Radioanalytical Chemistry,* Vol. 38, 1977, p. 267.
[*99*] Pouraghabagher, A. R. and Profio, A. E., *Analytical Chemistry,* Vol. 46, 1974, p. 1223.
[*100*] Orange, J. M. and Larson, J. G., *Analysis Instrumentation,* Vol. 13, 1975, p. 87.
[*101*] Lubkowitz, J. A. and Buenafama, H. D., *Analytical Chemistry,* Vol. 52, 1980, p. 233.
[*102*] Nadkarni, R. A., *Journal of Radioanalytical Chemistry,* Vol. 84, 1984, p. 67.
[*103*] Carter, J. A., Walker, R. L., and Sites, J. R. in *Trace Elements in Fuel,* S. P. Babu, Ed., Adv. Chem. Series No. 141, 1975, p. 74.
[*104*] Harding, A. R., Leland, D. J., and Leyden, D. J., paper presented to 14th Ann. Mtg. of FACSS, Detroit, Mich., 1987.
[*105*] Janssens, K. H. and Adams, F. C., *Journal of Analytical and Atomic Spectrometry,* Vol. 4, 1989, p. 123.
[*106*] Bilbrey, D. B., Leland, D. J., Leyden, D. E., Worbrauschek, P., and Aiginger, H., *X-Ray Spectrometry,* Vol. 16, 1987, p. 161.
[*107*] Smith, A. J., Rice, J. O., Shaner, W. C., and Cerato, C. C., *Amer. Chem. Soc., Div. of Petroleum Chem., Preprints,* Vol. 18, 1973, p. 609.
[*108*] Masour, S., Abu-Dagga, F., and Sabri, R., *Advances in X-Ray Analysis,* Vol. 29, 1985, p. 511.
[*109*] Christensen, L. H. and Agerbo, A., *Analytical Chemistry,* Vol. 53, 1981, p. 1788.
[*110*] Marangoni, C., Lutrario, P., and Tronca, A., *Metallurgica Italiana,* Vol. 64, 1972, p. 381.
[*111*] Larson, J. A., Short, M. A., Bonfiglio, S., and Allie, W., *X-Ray Spectrometry,* Vol. 3, 1974, p. 125.
[*112*] Kubo, H., Bernthal, R., and Wildeman, T. R., *Analytical Chemistry,* Vol. 50, 1978, p. 899.
[*113*] Krishnan, T. V., *Indian Journal of Pure and Applied Physics,* Vol. 15, 1977, p. 345.
[*114*] Denoyer, E. R. and Siegel, L. A., *Analytica Chimica Acta,* Vol. 192, 1987, p. 362.
[*115*] Yousif, A. N. and Al-Shahristani, H., *International Journal of Applied Radiation and Isotopes,* Vol. 28, 1977, p. 759.
[*116*] Stehr, K. C., *Advances in X-Ray Analysis,* Vol. 25, 1981, p. 173.
[*117*] West, H. M., *Advances in X-Ray Analysis,* Vol. 25, 1981, p. 177.
[*118*] Meier, H. and Unger, J., *Journal of Radioanalytical Chemistry,* Vol. 32, 1976, p. 413.
[*119*] Maddox, W. E. and Kelliher, W. C., *Advances in X-Ray Analysis,* Vol. 29, 1985, p. 497.
[*120*] Liu, Y. W., Harding, A. R., and Leyden, D. E., *Advances in X-Ray Analysis,* Vol. 29, 1985, p. 503.
[*121*] Dos Santos, J. M. F., Conde, C. A. N., Ferreira, L. F. R., and De Campos, A., *Advances in X-Ray Analysis,* Vol. 30, 1987, p. 237.
[*122*] Oxford Analytical Technical Note No. 85030.
[*123*] Oxford Analytical Technical Note No. 85033.
[*124*] Johnson, G., Kalnicky, D., Wallendorf, B., and Lass, B., *American Laboratory,* Vol. 20, No. 8, 1988, p. 58.
[*125*] Johnson, G., Kalnicky, D., and Lass, B., *P/PM Technology,* No. 7/8, 1988.
[*126*] Takahashi, Y., and Rey, M., *American Laboratory,* Vol. 15, No. 11, 1983, p. 27.
[*127*] Hieftje, G. M., *Journal of Analytical and Atomic Spectrometry,* Vol. 4, 1989, p. 117.

J. R. Mackey,[1] S. T. Watt,[1] C. A. Cardy,[1] S. I. Smith,[1] and C. A. Meunier[1]

Analysis of Additive Metals in Lubricating Oils

REFERENCE: Mackey, J. R., Watt, S. T., Cardy, C. A., Smith, S. I., and Meunier, C. A., **"Analysis of Additive Metals in Lubricating Oils,"** *Modern Instrumental Methods of Elemental Analysis of Petroleum Products and Lubricants, ASTM STP 1109*, R. A. Nadkarni, Ed., American Society for Testing and Materials, Philadelphia, 1991, pp. 52–61.

ABSTRACT: Additive elements such as Ba, Ca, Zn, Cu, Cl, P and Mg are added to lubricating oils for improved engine performance. Although necessary, additive concentrates are very expensive and it is important that refineries blend within the required specifications using reliable and accurate analytical techniques to measure the metal concentrations.

This paper is an overview of additive element analysis in lubricating oils at the Esso Petroleum Canada Research Centre. A comparison of the two current techniques, wavelength dispersive X-ray fluorescence (WXRF) and simultaneous inductively coupled plasma emission spectroscopy (ICPES), will be described.

We will give a brief summary on the scope, theory, and methodology of each technique plus the strengths and weaknesses of each method as they relate to fresh and used oils. A comparison of the precision between WXRF and ICPES will be examined as well as the use of robotics to improve precision and turnaround time with the ICPES method.

Our hope is to give the nonexpert a better understanding of the advantages and disadvantages of WXRF versus ICPES when choosing instrumentation or methodology for analyzing additive elements in fresh and used lubricating oils.

KEY WORDS: additive metals, lubricating oils, wavelength dispersive X-ray fluorescence, inductively coupled plasma emission spectroscopy

Additive concentrates are blended into base stocks to formulate finished lubricating oils which provide improved engine performance. Since these concentrates are very expensive, refinery labs require rapid, reliable, and accurate analysis to guarantee they have blended within specification targets.

Wavelength dispersive X-ray fluorescence (WXRF) has been used since the early 1960s [*1-6*] to monitor additive metals in lubricating oils. Long term precision of between 1 and 2% coefficient of variance at typical additive level concentrates has been the norm at Esso Petroleum Canada. The exception is magnesium (~7%) due to the longer magnesium X-ray radiation being absorbed before reaching the detector. The introduction of inductively coupled plasma emission spectroscopy (ICPES) in the mid-1970s [*7,8*] provided an alternative excitation source to WXRF. Although the precision for magnesium determination improved considerably with ICPES, the other elements showed poorer precision than WXRF, mainly due to manual weighing and dilution errors.

Recent developments in laboratory robotics [*9-11*] have totally automated both the sample preparation and the sample insertion into an ICPES. Robotic sample preparation techniques

[1] Esso Petroleum Canada, Research Department, P.O. Box 3022, Sarnia, Ontario, Canada N7T 7M1.

have provided ICPES methodology with advantages such as improved precision because of the reproducibility of the solvent dilution and unattended sample preparation.

This paper compares the current techniques, strengths, weaknesses, and precision for analyzing fresh and used lubricating oils at the Esso Petroleum Canada Research Centre.

Wavelength Dispersive X-Ray Fluorescence (WXRF)

Instrumentation

The instrumentation was as follows:

- Philips PW 1400 Sequential X-Ray Spectrometer System (50 to 60 Hz)
- X-ray Generator (60 kV, 4 kW)
- Chromium Anode Tube (3 kW)
- Analyzing Crystals
  - –LiF 200
  - –GE
  - –PX-1
- PDP 11/73 Minicomputer (Digital)

Principle of Operation

An oil sample is bombarded by high intensity X-rays. Each element present absorbs the X-rays and re-emits a characteristic spectrum. The secondary fluorescence is dispersed according to wavelength by a diffracting crystal. The intensity of the dispersed radiation is converted to electrical pulses and measured sequentially by a detector (Fig. 1). Detection

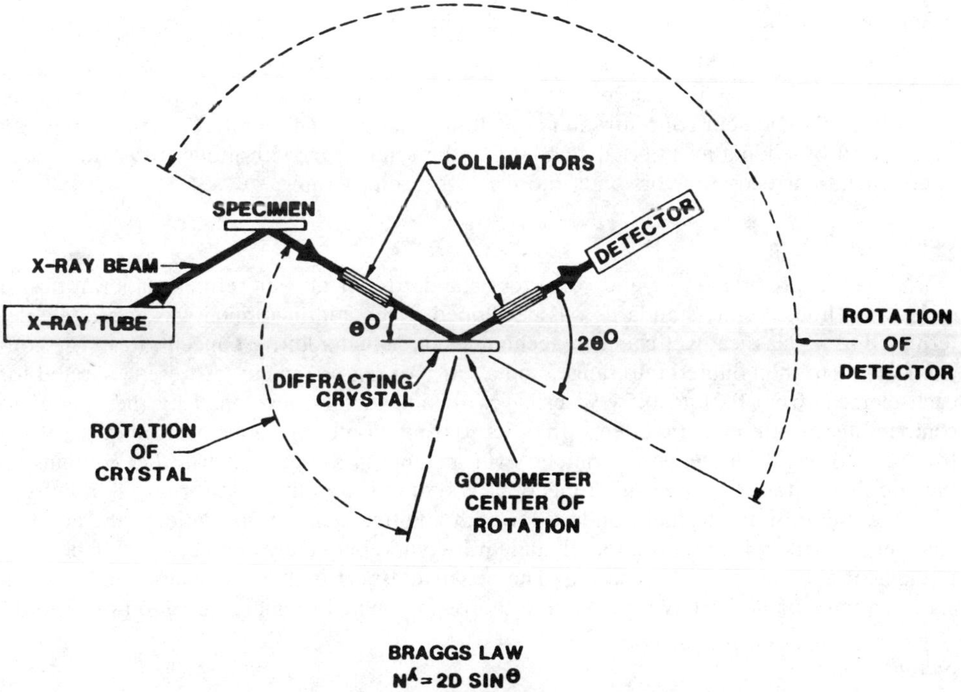

FIG. 1—*Optical arrangement of X-ray spectrometer.*

limits vary depending on the instrumentation, analyzing crystal, element, and matrix (density and other elements present).

Interferences

Matrix Effects—The depth that the X-rays penetrate the sample is dependent on the sample density. Higher density samples will cause less penetration and lower element intensity counts. These effects can be resolved by closely matching the density of the sample with that of the standard. Another technique is to add an internal standard to the sample. Any enhancement or suppression of the analyte intensity signal will similarly effect the internal standard intensity. Thus, by ratioing the analyte intensity over the internal standard intensity, these effects will be compensated.

Background Scattering—The secondary fluorescence is dispersed by a crystal monochromator according to Bragg's Law:

$$n\lambda = 2d \sin \theta$$

where

n = integer indicating the order of the spectrum,
λ = characteristic wavelength of element,
d = crystal interplaner spacing, and
θ = angle of incident beam to plane.

Here n represents the orders of reflection of any one wavelength that can interfere with another wavelength. Fortunately, these reflection orders can be filtered out by using a pulse height analyzer that discriminates between the different wavelength pulses.

Interelement Effects—Each element in the sample absorbs a portion of the initial X-ray radiation before re-emitting secondary fluorescence. The amount of radiation absorbed depends on the element concentration and atomic number. These interelement effects can be resolved by adding an internal standard to the sample or, with modern systems, using interelement correction factors built into the software program.

Calibration

In order to provide stable and consistent standards for all our refinery laboratories, a series of in-house X-ray standards was developed. Each additive metal was accurately determined using classical wet chemical techniques on actual additive concentrates. The concentrates were then diluted into range, and a series of regression curves were generated for each element from 0.01 to 0.5wt% using correction factors developed by the vendor to compensate for interelement effects. In order to compensate for slight spectrometer drifting from day to day each element is ratioed over a stable glass monitor disk that contains all the additive metals. Calibration on the WXRF system is usually good for 6 to 8 months.

The addition of an internal standard for each additive element and ratioing the additive element over its appropriate internal standard would adequately compensate for both interelement and spectrometer drifting. The negative aspect is that the sample and internal standard must be accurately weighed and thoroughly mixed to guarantee good precision.

Sample Analysis

Samples are analyzed as received and then correction factors applied to compensate for interelement effects. The intensities are ratioed over a monitor disk to adjust for spectrometer drift. A quality control (QC) sample is analyzed before each set of analyses to maintain precision control. The WXRF is also interfaced with an autosampler for unattended or overnight operation.

Inductively Coupled Plasma Emission Spectroscopy (ICPES)

Instrumentation

The instrumentation was as follows:

- Jarrell-Ash Atom Comp Series 1100 Direct Reading Spectrometer
- RF Generator (2.5 kW, 27.12 MHz)
- 2400 Grating Mirror (Resolution 0.036 nm)
- 0.75 Focal Plane (190 to 500 nm)
- 29 Element Array
- Dec PDP 1103 Computer

Principle of Operation

The oil sample diluted 1:50 with xylene is aspirated into the plasma where the elements are raised to their excited state in an energy zone equivalent to 7000 to 10000°C. Each element emits characteristic radiation which is collimated and directed onto a diffraction grating where the light from the elements are dispersed according to characteristic wavelengths. The dispersed energy intensities are simultaneously measured by a series of photomultiplier tubes (Fig. 2). ICPES instruments with air path monochromators are unable to analyze for chlorine or sulfur due to their absorbance in air at the low UV wavelengths.

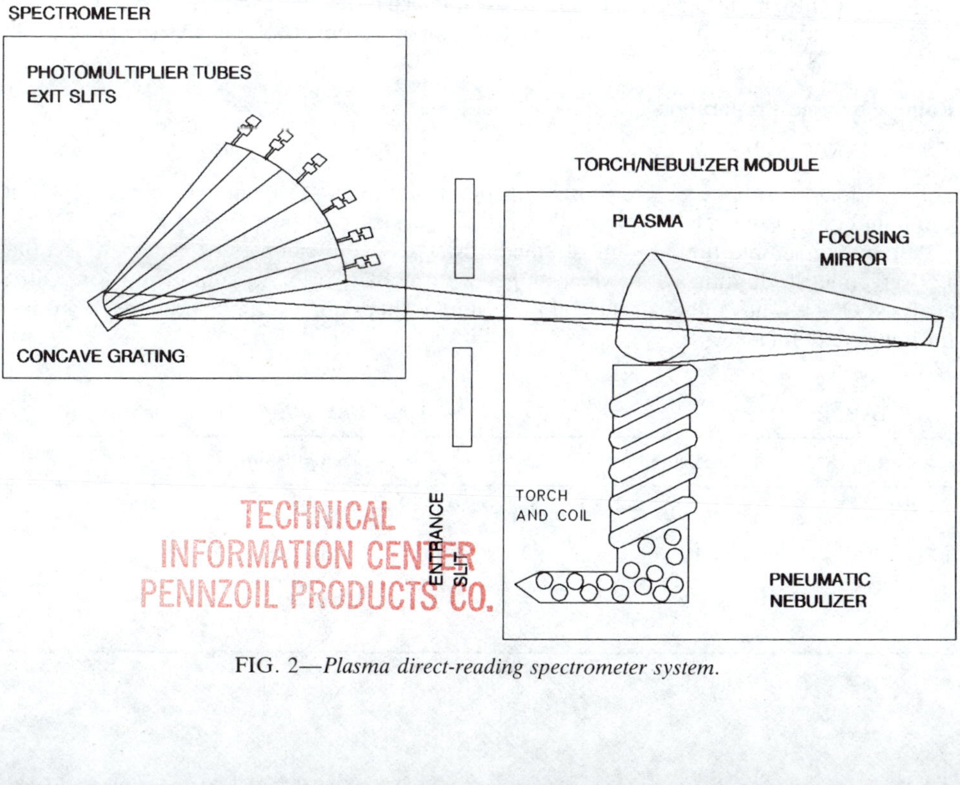

FIG. 2—*Plasma direct-reading spectrometer system.*

Vacuum path systems have the capability to analyze for sulphur in the UV range, but not chlorine. Detection limits for most elements are in the low ppm range.

Interferences

Matrix Effects—In ICPES methodology, the aspiration rate of the standard and sample solution into the spray chamber must be relatively equal or compensated for by use of an internal standard. Since the samples are initially diluted 50-fold with xylene, this guarantees that a xylene matrix will resolve most of the viscosity concerns. As suggested above, the use of an internal standard such as cobalt and ratioing all the additive metals over the cobalt intensity is a further safeguard against slight differences in viscosity and instrument drift. It is very important that samples are well characterized or screened in order that none of the selected internal standard is present in the sample.

Interelement Effects—It is possible to have a partial or direct overlap of one spectral line onto another, causing poor accuracy and precision. Interelement correction factors can be determined and added to the software package. If the interference is too severe, another wavelength line, even though less sensitive, should be used. In our present formulation packages, we have not found any spectral interferences. Table 1 lists the present wavelengths that are used for our additive metal procedure at Esso Petroleum Canada.

Calibration and Analysis

Multi-element oil standards are available for ICPES units from several vendors. A regression curve, ranging from 5 to 5000 μg/g is generated for each element. The multielement standards and samples are diluted (1:50) with xylene doped with cobalt naphthenate. The results are ratioed over the cobalt to compensate for slight variations in aspiration rates. The operator also monitors the cobalt intensity for each individual sample, for system failure. A QC or standard is analyzed after every ten samples to ensure the analysis is in control.

Robotic Sample Preparation

Zymate II Robot Operation

The Zymate II robotic system consists of stations for mixing, weighing, capping, pipetting and liquid dispensing. The layout of the robotic workcell is shown in Fig. 3.

Automated sample preparation of standards, QC's, and samples provides an accurate 1:50 gravimetric dilution with xylene doped with cobalt, prior to submission for ICPES analysis. One key advantage of the robotic system is that samples can be prepared overnight for improved efficiency.

TABLE 1—*Wavelength channels used on Jarrell-Ash ICPES at Esso Petroleum Canada.*

Element	Wavelength, nm
Ba	455.4
Ca	317.9
Zn	206.2
Cu	324.7
P	214.9
Mg	279.5

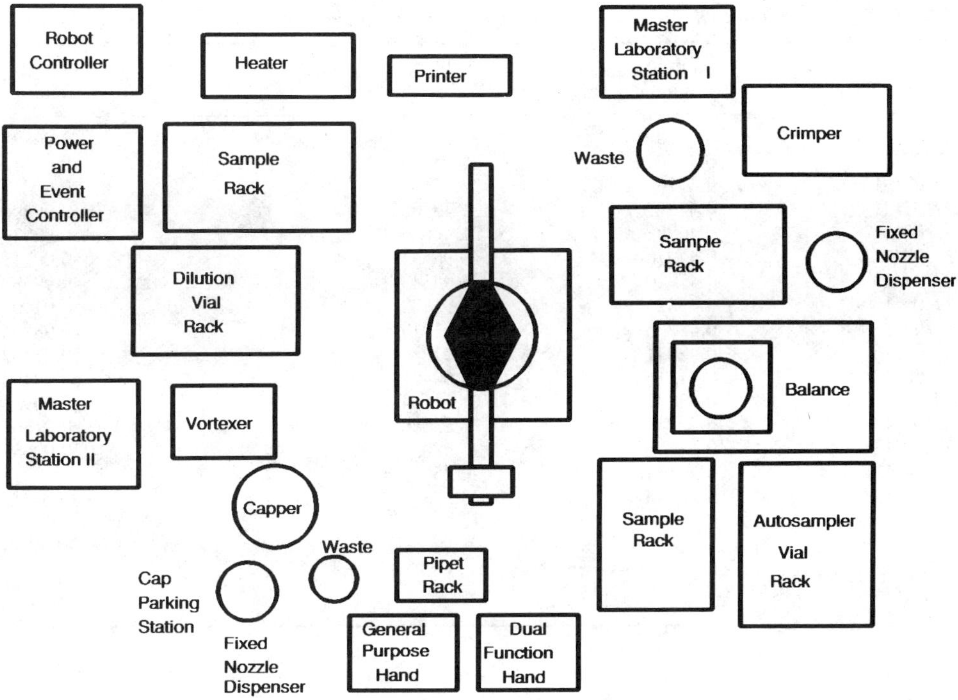

FIG. 3—*Schematic of Zymate II robotic system.*

Accuracy of Robot Dilution

During the development of a robotic sample preparation method for ICPES analysis, it became apparent that accurate dilutions for standards and samples would be necessary in order to maintain overall precision. Table 2 illustrates that robotic preparation over a period of two weeks obtained excellent repeatability (percent coefficient of variance of <0.1) on the multielement standard that was diluted 1:50 with xylene doped with cobalt.

Experimental Results and Discussion

Comparison Precision

Table 3 compares the percent coefficient of variance (repeatability) of direct WXRF, manual ICPES, and robotic ICPES. When comparing direct WXRF with manual ICPES, the repeatability of direct WXRF is better for all elements except magnesium. This poorer WXRF magnesium precision is consistent and predictable with atomic theory principles which state that as the atomic weight of an element decreases, the sensitivity of detection of the element also decreases. Thus, if better precision for magnesium is required, an ICPES technique would be recommended. As discussed earlier, Esso Petroleum Canada is unable to detect sulfur or chlorine with its present ICPES instrumentation.

Table 3 further demonstrates that a robotic ICPES technique can provide significant improvements in precision over a manual ICPES technique. In most cases the precision of the robotic/ICPES is comparable to direct WXRF, and in the case of magnesium the ICPES technique is much better.

TABLE 2—*Robot provides precise dilution ratios of multielement standard over two week period.*

Date Prepared	Sample Weight, g	Solvent Weight, g	Dilution Ratio
87/07/31	0.43250	21.62466	49.99933
87/07/31	0.43861	21.95075	50.04615
87/07/31	0.44153	22.07664	50.00029
87/07/31	0.42589	21.29282	49.99606
87/08/04	0.43788	21.90685	50.02939
87/08/12	0.47970	23.98554	50.00111
87/08/13	0.47910	23.95343	49.99680
87/08/13	0.43825	21.91732	50.01104
87/08/13	0.42268	21.12919	49.98863
87/08/13	0.42047	21.02841	50.01152
87/08/13	0.47216	23.60421	49.99193
87/08/13	0.42222	21.11504	50.00954
87/08/13	0.42248	21.13086	50.01605
87/08/13	0.41155	20.58036	50.00688
87/08/13	0.40522	20.25627	49.98832
87/08/13	0.46540	23.27533	50.01138
			$\bar{x}$ = 50.00653
SD = 0.01873			
% CV = 0.03746			

Comparison of Additive Metal Procedures

Table 4 compares the strengths and weaknesses of each of the additive metal procedures used at Esso Petroleum Canada. The important parameters are discussed in further detail below. Precision and interferences will not be discussed, since they were addressed earlier.

Scope—WXRF can analyze for all eight additive elements. The concentration range is from 100 to 5000 ug/g except for magnesium with a significantly poorer detection limit of 500.

Our present air-path ICPES spectrometer cannot detect sulfur or chlorine. Vacuum path systems could analyze for sulphur in the UV range, but not chlorine. The concentration range for the six detectable elements is 5 to 5000 ug/g.

TABLE 3—*Comparison of precision of direct XRF versus manual ICPES versus robotic ICPES.*

		% Coefficient of Variance		
Metal	Typical Wt%	Direct XRF	Manual ICPES	Robotic ICPES
Zn	0.16	0.9	1.7	1.4
Cu	0.01	1.7	2.4	1.8
Ba	0.51	0.7	1.3	1.1
Ca	0.21	0.6	1.1	0.7
Cl	0.14	1.0	...[a]	...[a]
S	0.20	1.2	...[a]	...[a]
P	0.16	1.8	2.6	1.9
Mg	0.10	7.5	1.2	0.9

[a] ICPES cannot analyze Cl or S.

TABLE 4—*Comparison of additive metal procedures.*

Parameter	Direct WXRF	Manual ICPES	Robotic ICPES
• Scope	• All additive elements • 100 to 5000 ppm (Mg = 500 to 5000 ppm)	• Cannot do S or C1 • 5 to 5000 ppm	• Cannot do S or Cl • 5 to 5000 ppm
• Calibration	• Every 6 months	• Daily	• Daily
• Sample preparation	• Minimal	• Labor intensive	• Minimal, but robot intensive
• Instrument analysis	• Sequential • Nondestructive	• Rapid, simultaneous • Destructive	• Rapid, simultaneous • Destructive
• Total analysis time	• 20 min • Overnight analysis	• 15 min • Labor intensive	• 30 min • Robot intensive
• Precision	• Good, except for light element (Mg)	• Poorer than XRF, but much better for Mg	• Better than manual ICPES • Comparable to XRF, but much better for Mg
• Interferences	• Corrected by interelement correction factors • Additional elements require development of new correction factors	• Not significant • Additional elements easy to correct for	• Same as manual ICPES
• Fresh oil analysis	• Excellent (industry standard)	• Good	• Very good
• Used oil analysis	• Poor, particulates interfere	• Best direct technique available • Larger particle sizes can interfere	• Same as manual ICPES
• Cost	• $380 K including autosampler	• $150 K including autosampler	• $210 K including autosampler and robot

Calibration—State-of-the-art WXRF instruments usually require calibration about once every six months. Although calibration is infrequent, the process will take as much as one day. ICPES techniques require daily calibration, but calibration can be completed in about 15 to 20 min.

Sample Preparation and Analysis—Sample preparation for WXRF analysis requires minimal operator time. The WXRF sample cup is introduced directly into the unit with no sample preparation required.

Sample preparation for the manual ICPES method is very labor intensive and requires the accurate weighing and dilution of standards and samples before introduction to the unit. Sample preparation for the robotic ICPES method also requires accurate weighings and dilutions, but this is robot intensive and can be performed on off-shifts.

The WXRF analysis is a sequential, nondestructive technique requiring about 15 to 20 min to analyze all eight additive elements in a sample. Fortunately this time is not labor intensive and can be performed overnight using an autosampler.

The ICPES analysis is a simultaneous, destructive technique requiring approximately 2 to 3 min per sample for six elements. An ICPES technique is not suited to unattended operation due to possible problems with the plasma and the sample delivery. Although vacuum path systems can analyze for sulfur, analysis of chlorine remains impossible due to volatility and memory effects.

Fresh versus Used Oil Analysis—Newly formulated (fresh) lubricating oils are easily analyzed by both WXRF and ICPES techniques, because all the elements are in solution and the samples can be diluted and/or spiked with a suitable internal standard.

Used oil analysis has many associated problems for both techniques. Particulates may interfere with the WXRF method, resulting in poor precision. Since X-rays only penetrate to a depth of a few micrometers into the sample, any particulates that migrate to the film surface will either enhance or suppress the signal depending on the physical state of the element in the sample. ICPES techniques can overcome much of the problem by keeping the finely dispersed particles continually agitated prior to aspiration of the solution into the unit. Unfortunately, as the particle size increases, the aspiration source (nebulizer) will restrict the larger particles, resulting in lower element recoveries. This situation is difficult to predict and almost impossible to resolve without using alternative sample preparation techniques such as long and tedious wet acid digestions.

Instrument Costs—Overall instrument costs can be an issue. Most state-of-the-art WXRF spectrometers including an autosampler will cost about $380 K. A simultaneous ICPES unit with 20 element channels and an autosampler is significantly less expensive at about $150 K. Further automation of the ICPES sample preparation using a robot will cost an additional $60 K.

Summary

The choice of instrumentation and methodology to analyze for additive metals in lubricating oils can vary for different situations and laboratories. It is hoped that the nonexpert has been given a better understanding of the strengths and weaknesses of WXRF and ICPES, which are both presently employed at the Esso Petroleum Canada Research Centre. It is important before choosing an instrument appropriate to each laboratory and application that criteria of "musts" and "wants" be carefully established. We hope the reader can then reach a logical decision based on matching his criteria with the most suitable instrumentation to meet his needs.

References

[1] Bertin, E. P., *Principles and Practice of X-Ray Spectrometric Analysis*, Plenum Press, New York, 1975.

[2] Claisse, F. and Tertian, R., *Principles of Quantitative X-Ray Fluorescence Analysis,* Heyden and Son Ltd., London, 1982.

[3] Barrett, C. S. et al., *Advances in X-Ray Analysis*, Vol. 32, Plenum Press, New York, 1989.
[4] Bertin, E. P., *Introduction to X-Ray Spectrometric Analysis,* Plenum Press, New York, 1985.
[5] Jenkins, R., Gould, R. W., and Gedcke, D., *Quantitative X-Ray Spectrometry*, Marcel Dekker Inc., New York, 1981.
[6] Jenkins, R. and DeVries, J. L., *Worked Examples in X-Ray Analysis,* Springer-Verlag, New York, 1972.
[7] Parsons, M. L., Forester, A., and Anderson, D., *An Atlas of Spectral Interferences in ICP Spectroscopy,* 1980.
[8] Barnes, R. M., *Applications of ICP to Emission Spectroscopy,* 1977.
[9] Jones, R. D. and Cross, J. B., "Automating Sample Preparation and Disposal with a Robotic Workcell," in *Advances in Laboratory Automation-Robotics, 1985*, G. L. Hawk and J. R. Strimaitis, Eds., Zymark Corporation, Hopkinton, Mass., 1985, p. 293.
[10] Becker, D. A., "Robot Control of ICP Analysis of Lubricating Oils," in *Advances in Laboratory Automation-Robotics*, *1984*, G. L. Hawk and J. R. Strimaitis, Eds., Zymark Corporation, Hopkinton, Mass., 1984, p. 35.
[11] Jones, R. D., Meyer, M. E., and Cross, J. B., "Developing a Flexible Intelligent Robotized ICP System," in *Advances in Laboratory Automation-Robotics*, Vol. 5, G. L. Hawk and J. R. Strimaitis, Eds., Zymark Corporation, Hopkinton, Mass., 1989, p. 167.

Michael Gonzales[1] *and Arthur W. Lynch*[1]

Methods Development for Trace Metal Characterization of Crude Oil by Inductively Coupled Plasma Atomic Emission Spectroscopy

REFERENCE: Gonzales, M. and Lynch, A. W., **"Methods Development for Trace Metal Characterization of Crude Oil by Inductively Coupled Plasma Atomic Emission Spectroscopy,"** *Modern Instrumental Methods of Elemental Analysis of Petroleum Products and Lubricants, ASTM STP 1109,* R. A. Nadkarni, Ed., American Society for Testing and Materials, Philadelphia, 1991, pp. 62–69.

ABSTRACT: Metals and other contaminating elements found in low concentrations (10 ppm or less) in crude oil can alter the conversion of refinery feedstocks by poisoning catalysts. Organic matrices, such as crude oil, can cause complications when analyzing for trace elements by inductively coupled plasma atomic emission spectroscopy (ICP-AES). This study addresses some of the problems associated with direct analysis of P, Mo, Zn, Pb, Ni, Sn, Fe, Cr, V, Ca, Cu, Ti, Ba, As, Hg, Se, Sb, and Mg in crude oil using this technique. Specific areas addressed were solvent selectivity, spectral overlap, and the use of an internal standard to overcome sample inconsistencies such as viscosity. Kerosine was selected, instead of xylene or MIBK, as the best solvent based on a study of detection limits. A list of preferred wavelengths, included in the report, is based upon a scanning study that determined spectral overlap and sensitivity. The use of scandium as an internal standard overcomes delivery and nebulization problems associated with variations in viscosity and also drift in the instrument. Recovery experiments on crude oil samples containing known amounts of the elements of interest show greater than 90% recovery on 14 of the 18 elements listed.

KEY WORDS: crude oil, crude oil analysis, trace metals, ICP-AES, internal standardization, kerosine, viscosity

The determination of metals content in crude oil is an important part of any crude oil quality program. Even though the oil is typically desalted prior to distillation, small amounts (<1 ppm) of some elements (i.e. Pb, As, Hg) remaining in the oil can rapidly deactivate some refinery catalysts [*1*] or end up in the product. In addition, the potential adverse environmental impacts of metals such as these require knowledge of the metals content in refinery effluents and waste water [*2*]. Therefore a determination of the metal concentration in crude oil before and after processes such as desalting is of value to refinery operations. Inductively coupled plasma atomic emission spectroscopy (ICP-AES) has become one of the primary methods for determining the concentration of metals in aqueous solutions because of its high sensitivity and the short time required for multielement analysis. The U.S. Environmental Protection Agency Contract Laboratory Program (EPA/CLP) [*3*] contains recommended procedures for ICP-AES analysis of ground water, but the procedures

[1] Sandia National Laboratories, Albuquerque, NM 87185.

do not include all the elements of interest in crude oil. We have included P, Ti, Sn, and Mo because of their potential impact on refinery operations.

Historically, the indirect analysis of nonaqueous samples required sample preparation techniques such as dry or wet ashing to concentrate inorganic constituents and make them soluble in aqueous solution. The drawbacks to these (indirect) methods are that they can be time consuming and there may be analyte loss due to volatility of some elements during ashing [*4*]. This study was done to determine if the sensitivity and accuracy of the ICP analysis could be maintained while using direct injection of diluted oil.

The first part of this study included compilation of a list of elements that could be of environmental and operational concern to refinery operators. The list contains not only naturally occurring metals normally considered to be major catalyst or product contaminants or both (i.e., Cu, V, Ni, Fe), but also contains those elements that may enter the system as contamination from a source such as an inadequately cleaned tanker used to transport the oil. The spectral wavelengths used for analyses of these selected elements were chosen from among many candidates by incorporating a scanning technique that identified any cross-element interferences. We were also able to monitor the sensitivity, peak profiles, and baseline resolution of these candidate wavelengths. In some instances, the wavelengths selected differed from those recommended by EPA/CLP due to interferences identified by the scanning technique.

In addition to identifying instrumental parameters, other factors, such as the effects of viscosity and diluents, were addressed. Differences in viscosity after a 1 to 5 dilution (factors of 2 or 3 are not uncommon) cause changes in atomization behavior and therefore can have significant effects on results. The effects of several diluents were compared, and the results showed that kerosine provided better detection limits. To further mitigate the potential problems associated with viscosity, the use of scandium as an internal standard was investigated.

Equipment

Oil analyses were done on an ARL Model 3580 combination simultaneous/sequential inductively coupled plasma atomic emission spectrometer (ICP-AES). The instrument contains a single argon plasma source connected to a direct reading 1-m polychromater for simultaneous multielement analyses and a 1-m scanning monochromater for sequential analyses. The detectors are capable of operating simultaneously and are controlled by a DEC micro PDP-11 with a 10 Mbyte Winchester disk. Both the monochromater and the polychromater are under vacuum to enhance sensitivity in the ultraviolet range. Samples were introduced into the plasma torch by a Meinhard Type A concentric glass nebulizer connected to a conical spray chamber. Table 1 contains the operating conditions used in this study.

Selection of Parameters

Wavelength Selection

The spectral wavelengths of the elements listed in Table 2 were selected on the basis of sensitivity and their lack of spectral overlap. Interelement interference is just as important a concern as sensitivity. The first indications of spectral interferences are biases that sometimes show up as nonlinear calibration curves. Typically the interelement interferences will be integrated along with the intensity reading from the element of interest resulting in quantitative errors. Wavelength scans of all the individual elements were performed at each of the most common wavelength lines to determine which elements had overlapping in-

TABLE 1—*Operating conditions.*

RF Generator Readings	Value
Incident power	1400 W
Reflected power	11 W
Plate current (<0.75 A)	0.72
Filament voltage (7.1–7.9 V)	7.8
High voltage (<4800 V)	4200
Grid current (<0.4 A)	0.15
ICP Source Parameters	**Value**
Coolant argon meter reading (psi)	36
Plasma argon meter reading (psi)	35
Nebulizer argon meter reading (psi) (@ 1 L/min flow)	0.37
Nebulizer uptake rate (mL/min)	1.2–1.3

tensities. A total of 1276 scans (blanks, crude oil, solvents, and 26 individual elements) was run. All standard solutions were made from individual organometallic concentrates and a multielement concentrate, S-21 (Conostan, Ponca City, Oklahoma), containing 21 of the most common elements used in petroleum elemental analysis. Also included were scans of the elements contained in S-21 that were not included in the list compiled for this report (Al, Ag, B, Cd, Mn, Na, and Si). The elements were scanned in a base oil (Conostan #20 base oil)/kerosine matrix. The individual element solutions contained 100 μg/g. This concentration was chosen to insure a high intensity reading.

A total of 44 wavelength channels was examined. The region on either side of the peak for 27 wavelength channels was scanned by the simultaneous direct reading polychromater. For these, the scans covered a range of 0.03 nm on either side of the emission line. This proved to be a wide enough range to provide a peak profile without the added scanning

TABLE 2—*Selected wavelengths and detection limits.*

			Detection Limit, μg/g	
Element	Sim/Seq	Wavelength	Kerosine	Xylene
P	sim	178.29	0.0397	0.5834
Mo	sim	202.03	0.0116	0.3348
Zn	sim	213.86	0.0012	0.0056
Pb	sim	220.35	0.0256	0.4659
Ni	sim	231.60	0.0103	0.2756
Sn	sim	242.95	0.0602	0.1553
Fe	sim	259.94	0.0031	0.0074
Cr	sim	267.72	0.0044	0.0246
V	sim	309.31	0.0015	0.0062
Ca	sim	317.93	0.0019	0.0286
Cu	sim	324.75	0.0026	0.0017
Ti	sim	337.28	0.0015	0.0066
Ba	sim	493.41	0.0011	0.0035
As	sim	189.04	0.0583	0.5404
Hg	seq	184.95	0.0083	0.01077
Se	seq	196.09	0.0400	0.05630
Sb	seq	217.58	NA	NA
Mg	seq	285.21	0.0010	0.0033

NA = not analyzed.

time that a wider range would require. The remaining 17 wavelength channels were scanned using the sequential monochromater whose range covered 0.15 nm on either side of the emission line with a scan increment of 0.015 nm and a scan integration time of 0.5 s. In this paper our selected simultaneous wavelengths are listed by increasing wavelength for easy reference. The sequential channels are analyzed in increasing wavelength to minimize the travel time between wavelength changes on the sequential monochromater.

An example of spectral overlap is shown in Fig. 1. The scan data at the antimony wavelength of 206.83 nm clearly show a contribution from the vanadium solution. Whether or not the interference is from vanadium or from some other component in solution is not known. Iron has been shown to interfere with several elements in a variety of ICP-AES applications [5]. Iron can typically be found as a constituent in crude oil. Therefore an experiment was conducted to determine if there were any interferences, due to increasing Fe concentrations, with the intensities of the selected elements and scandium. The results in this study indicate that an Fe concentration of >1000 μg/g is required before significant interference occurs to the elements in our matrix.

Solvent Selection

Many crude oils cannot be aspirated directly due to their high viscosity. Solvents such as xylene, methyl isobutyl ketone (MIBK), and kerosine, which are ICP compatible and of low viscosity, dilute the solutions sufficiently to allow effective nebulization. Many analysts prefer not to use MIBK because it produces the least stable plasma [*6*]. During this work MIBK could not be used because it caused the plasma to continually extinguish. Kerosine has a high flash point and would probably be a relatively safe solvent to use. The ARL Model 3580 includes a software program for determining detection limits. It compares a blank to a known concentration and reports the concentration that is twice the standard deviation of the background emission at the selected wavelength. Xylene was compared with kerosine to determine which solvent provided the lowest detection limits for the elements of interest. Kerosine was selected as the best diluent based on the results in Table 2. Antimony was not included in the elemental matrix during this study. It is included in Table 2 only to indicate its preferred wavelength.

Internal Standard

The sensitivity of ICP-AES is directly related to nebulization efficiency and therefore viscosity. Even with dilution, the specimen viscosity remains variable due to different viscosities of crude oils. For this reason, the use of an internal standard to correct for the

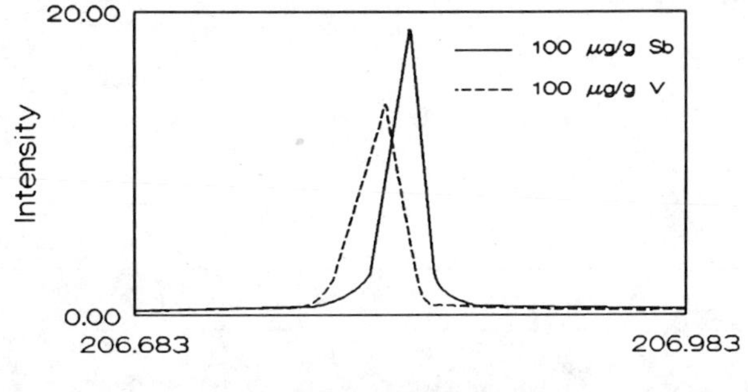

FIG. 1—*Spectral overlap of V solution on the Sb 206.83 nm line.*

differing flow rates of various oils [7] was necessary. We chose Sc as the internal standard because it satisfied the criteria for internal standards by being non-interfering, and it is not usually present in the samples being analyzed. The procedures for preparing the reagents, standards and samples listed below are based on the internal standard method. The 361.38 nm line was selected for Sc.

Analytical Procedure

Standards

Known concentrations of all elements of interest were obtained from Conostan as solutions in a base oil. Conostan S-21, obtained at a concentration of 500 μg/g each, was used to prepare the 100 and 10 μg/g standard reagents for this work. However, the S-21 Conostan mixture does not include all the elements of interest; the elements Sb, As, Hg, and Se were added to the S-21 or standardized separately. These elements were obtained as 5000 μg/g (Sb) and 100 μg/g (As, Hg, and Se) solutions. The manufacturer indicates that As, Hg, and Se standards are not stable in the long term when mixed as components of a large mixture but should be acceptable as a mix among themselves. Scandium was used as the internal standard and was obtained as a 2000 μg/g solution.

Prepared Reagents

The reagents (I-V) specified below were used to prepare the eight calibration standards listed in Table 3 and for diluting concentrated samples (Reagent V).

Reagent I: 100 μg/g of All Elements in S-21 and Sb

2.00 g	500 μg/g S-21
0.20 g	5000 μg/g Sb
7.80 g	base oil
10.00 g	@ 100 μg/g S-21 and Sb

TABLE 3—*Preparation of calibration standards.*

	Blank									
	I	II	A	B	C	D	E	F	G	H
					Quantity Added—grams					
Reagent I	0.0	0.0	0.0	0.0	0.0	2.5	5.0	0.0	0.0	0.0
Reagent II	0.0	0.0	0.5	2.5	5.0	0.0	0.0	0.0	0.0	0.0
Reagent III	0.0	0.0	0.0	0.0	0.0	0.0	0.0	2.5	5.0	10.0
Base Oil	10.0	10.0	9.5	7.5	5.0	7.5	5.0	7.5	5.0	0.0
Kerosine	40.0	0.0	0.0	0.0	0.0	0.0	0.0	0.0	0.0	0.0
Reagent IV	0.0	40.0	40.0	40.0	40.0	40.0	40.0	40.0	40.0	40.0
					Concentration in μg/g					
S-21	0.00	0.00	0.10	0.50	1.00	5.00	10.00	0.00	0.00	0.00
As, Hg, Se	0.00	0.00	0.00	0.00	0.00	0.00	0.00	1.65	3.30	6.60

Reagent II: 10 μg/g of All Elements in S-21 and Sb

1.00 g	Reagent I
9.00 g	base oil
10.00 g	@ 10 μg/g S-21 and Sb

Reagent III: 33 μg/g of As, Hg, and Se

6.66 g	100 μg/g As
6.66 g	100 μg/g Hg
6.66 g	100 μg/g Se
19.98 g	@ 33 μg/g As, Hg, and Se

Reagent IV: Scandium in Kerosine (Internal Standard)

2.50 g	2000 μg/g Sc
797.50 g	kerosine
800.00 g	@ 6.25 μg/g Sc in kerosine

Reagent V: Scandium in Kerosine (for Dilutions)

2.00 g	2000 μg/g Sc Conostan standard
158.00 g	base oil
640.00 g	kerosine
800.00 g	@ 5.00 μg/g Sc in kerosine

Calibration Standards

The calibration solutions described in Table 3 (approximately 50 g total weight) were used to calibrate and determine linearity and detection limits on the ICP-AES. All solutions were prepared by adding 1 part reagent to 4 parts kerosine (by weight) to minimize inconsistencies due to differences in viscosity. Each reagent was accurately weighed to the nearest 0.01 g (Table 3 contains ideal weights). Scandium, at a final concentration of 5 μg/g, was used as an internal standard for all samples and was added as Reagent IV - "scandium in kerosine". The concentration of each element in the final standard solution is calculated using the equation

$$\frac{\text{conc. of element} \times \text{grams of reagent}}{\text{grams of mixture after Reagent IV addition}} = \mu\text{g/g}$$

The use of the standardization matrix described in Table 3 enables the previously mentioned incompatible elements (As, Hg, and Se) to be included in a single standardization routine. Typically, separate standardization and analysis programs are run to include both sets of elements.

Calibration Matrix

Table 4 shows the resulting concentrations of each calibration solution. Once linearity is established for each element, a one or two point standard curve is sufficient for calibration during routine analysis. However, linearity should be re-established after any alteration of

TABLE 4—*Calibration data entries for analysis program.*

	Blank									
	I	II	A	B	C	D	E	F	G	H
P	0.00	0.00	0.10	0.50	1.00	5.00	10.00	x	x	x
Mo	0.00	0.00	0.10	0.50	1.00	5.00	10.00	x	x	x
Zn	0.00	0.00	0.10	0.50	1.00	5.00	10.00	x	x	x
Pb	0.00	0.00	0.10	0.50	1.00	5.00	10.00	x	x	x
Ni	0.00	0.00	0.10	0.50	1.00	5.00	10.00	x	x	x
Sn	0.00	0.00	0.10	0.50	1.00	5.00	10.00	x	x	x
Fe	0.00	0.00	0.10	0.50	1.00	5.00	10.00	x	x	x
Cr	0.00	0.00	0.10	0.50	1.00	5.00	10.00	x	x	x
V	0.00	0.00	0.10	0.50	1.00	5.00	10.00	x	x	x
Ca	0.00	0.00	0.10	0.50	1.00	5.00	10.00	x	x	x
Cu	0.00	0.00	0.10	0.50	1.00	5.00	10.00	x	x	x
Ti	0.00	0.00	0.10	0.50	1.00	5.00	10.00	x	x	x
Ba	0.00	0.00	0.10	0.50	1.00	5.00	10.00	x	x	x
As	0.00	0.00	x	x	x	x	x	1.65	3.30	6.60
Hg	0.00	0.00	x	x	x	x	x	1.65	3.30	6.60
Se	0.00	0.00	x	x	x	x	x	1.65	3.30	6.60
Sb	0.00	0.00	0.10	0.50	1.00	5.00	10.00	x	x	x
Mg	0.00	0.00	0.10	0.50	1.00	5.00	10.00	x	x	x
Sc	x	x	x	x	x	x	x	x	x	x

Note: Where an "x" is shown, no value is entered.

the instrument that could affect the instrument performance (i.e., replacement of parts such as photomultiplier tubes). If it is determined that a specimen contains one or more elements with concentrations outside the range of established calibration, the specimen must be diluted with Reagent V (5 μg/g Sc in kerosine) to lower the concentration of those elements to within the range of the calibration standards.

Analysis of Spiked Crude Oil

Analyses were performed on six crude oil samples spiked with approximately 10 μg/g of the elements in the S-21 standard and approximately 5 μg/g of As, Hg, and Se each. Ten gram specimens of the spiked oil were diluted with 40 g of Reagent IV, resulting in final concentrations of 2 μg/g for the S-21 elements and 1 μg/g for the As, Hg, and Se. The purpose was to determine the extent (if any) of matrix effects due to a raw crude oil. The percent recoveries were calculated by subtracting the concentration determined in an unspiked blank from the concentration found in the spiked specimen and then dividing by the concentration in the spike. The results are listed in Table 5.

The results of the crude oil matrix experiment show a recovery of 90% or greater for most of the elements. The exceptions are tin, arsenic, mercury, and selenium. The matrix effect on the arsenic was severe enough to totally mask the signal from the element. It might be noted that the Conostan As, Hg, and Se are known for their instability, and this fact may play a role in the lack of recovery for these elements. Regardless, alternative methods such as graphite furnace and vapor atomic absorption techniques can be used for their determination. Meanwhile, these elements can remain in the ICP scheme as a semiquantitative screening analysis with final concentration verification by the alternative methods. The presence of surplus elements from S-21 with the crude oil matrix was just one of the

TABLE 5—*Results from the spiked crude oil experiment.*

Element	% Recovery	Element	% Recovery	Element	% Recovery
P	93.1 ± 3.3	Fe	96.8 ± 1.6	Ba	102.4 ± 1.4
Mo	94.2 ± 1.8	Cr	96.1 ± 1.7	As	ND
Zn	93.5 ± 1.5	V	94.8 ± 1.3	Hg	54.5 ± 24.5
Pb	91.1 ± 1.9	Ca	95.9 ± 1.6	Se	49.9 ± 4.4
Ni	91.0 ± 1.2	Cu	101.5 ± 1.4	Sb	101.9 ± 2.6
Sn	72.9 ± 5.1	Ti	98.0 ± 1.6	Mg	102.6 ± 4.4

ND = not detected

foci of this study. The use of multielement standards certainly have their limitations, but they have their advantages as well. Further experiments with individual elements may shed some more light on the issue of element instability versus crude oil matrix effects.

Conclusions

1. The use of direct ICP analysis has overcome the main drawbacks of time consuming ashing procedures and potential loss of analyte due to volatilization. Unless some precautions have been taken to prevent the loss of analytes through the use of a containment bomb or reflux apparatus, volatility of analyte will be inherent in indirect methods. Combining an internal standard with an appropriate organic solvent has overcome drift, sample transport delivery, and nebulization problems associated with different sample viscosities and surface tension variations. The instrument corrects the analyte concentration by monitoring the internal standard, thus allowing us to maintain accuracy for most of our selected elements. Scandium, at a 5 μg/g concentration, was found to be an excellent internal standard in a crude oil matrix. We were able to determine 14 out of 18 elements with recoveries greater than 90%.

2. Kerosine was chosen as the diluent because it provided better detection limits than xylene for most of the elements in our matrix. The use of kerosine at a 1 to 5 dilution helps to normalize the various viscosities.

Acknowledgments

We wish to thank William B. Chambers and Sandra L. Chavez, Sandia National Laboratories, for their cooperation and support of this project.

References

[*1*] Nelson, W. L., *Petroleum Refinery Engineering*, Fourth ed., McGraw-Hill, New York, 1958, p. 265.
[*2*] Nelson, W. L., *Petroleum Refinery Engineering*, Fourth ed., McGraw-Hill, New York, 1958, pp. 268-269.
[*3*] USEPA-Contract Laboratory Program, Statement of Work for Inorganic Analysis, No. 788, Nov. 1986.
[*4*] Sharma, S. N., *Journal of Science and Industrial Research*, Vol. 42, 1983, p. 341.
[*5*] Parsons, M. L. and Forster, A., *Atlas of Spectral Interferences in ICP Spectroscopy*, Plenum Press, New York and London, 1980.
[*6*] Nygaard, D. D., Chase, D. S., and Leighty, D. A., "Analysis of Hydrocarbon Oils," Instrumentation Laboratory Applications Note, AID #166, Wilmington, Mass.
[*7*] Schmidt, G. J. and Slavin, W., *Analytical Chemistry*, Vol. 54, 1982, p. 2491.

J. M. Carter,[1] W. Batie,[1] and A. E. Bernhard[1]

Automated Rapid Multielement Analysis of Oils for Wear Metals by Flame Atomic Absorption Using a Single Aspiration Per Sample

REFERENCE: Carter, J. M., Batie, W., and Bernhard, A. E., **"Automated Rapid Multielement Analysis of Oils for Wear Metals by Flame Atomic Absorption Using a Single Aspiration Per Sample,"** *Modern Instrumental Methods of Elemental Analysis of Petroleum Products and Lubricants, ASTM STP 1109,* R. A. Nadkarni, Ed., American Society for Testing and Materials, Philadelphia, 1991, pp. 70–76.

ABSTRACT: In recent years, atomic absorption (AA) has been abandoned by some laboratories as the technique of choice for wear metal analysis in favor of other techniques. Although AA has been generally regarded as a good method for this type of analysis, economic considerations seem to have dictated this trend. Conventional AA analysis is too slow and tedious for multielement analyses. Laboratories need faster turnaround times and less labor intensive methods such as inductively coupled plasma (ICP) and spark emission, both of which are better suited to multielement analysis. A new sequential atomic absorption spectrophotometer, capable of rapid multielement analysis for up to 24 elements per sample in a single run, has been evaluated for oil analysis. The instrument is able to change hollow cathode lamps and other instrumental parameters automatically within 2 s, thus allowing 10 elements to be determined per minute. Unlike conventional AA, a sample is completely analyzed for all elements before the following sample is aspirated.

KEY WORDS: wear metals, multielement analysis, oil analysis, atomic absorption

During the last few years, many laboratories involved in the analysis of oil for wear metals have replaced or supplemented their use of atomic absorption spectrophotometers (AA) by adding inductively coupled plasma (ICP) or spark emission (rotrode) instruments. This change may have come about reluctantly, because atomic absorption has been shown to be a useful way of providing good results for this type of analysis. The main reason for the change is a matter of economics. Conventional AAs cannot really determine all the elements in one sample rapidly; therefore the turnaround time per sample is prohibitive. Conventional AA analysis is also very labor intensive, in an age where more automated instruments are desired. If a conventional automated AA spectrophotometer is used to analyze 100 samples for 10 elements, it determines one element in all 100 samples in sequence, then switches to the second element revisiting all 100 samples, and so on. Thus it does not complete the analysis for the first sample until it has measured 9 elements in all 100 samples. Even the most automated, computer-controlled AAs merely mimic manual operation. In order to run this many elements, a great deal of changes must be made by the operator, either human

[1] Analyte Corporation, Medford, OR 97504.

or computer: hollow cathode lamps, the analytical wavelength, slit width, and flame conditions all must be changed between elements. In comparison, the appeal of ICP or spark emission is that the multielement analysis of a sample is completed for all 10 elements before moving on to the next sample. Therefore the analysis is completed more quickly.

A new multielement AA has been developed which, unlike conventional AA instruments, analyzes one sample completely for all elements before starting the next sample. It is able to make determinations at the rate of 10 elements per minute, thus making it faster than sequential ICP and spark emission.

Experimental Procedure

The new sequential mode AA (ANALYTE 16) has a 16-position hollow cathode lamp turret that is capable of selecting a new lamp within 2 s. By using multielement lamps up to 24 elements can be measured in a single run. During the 2 s interval that the lamp is being changed, the instrument also adjusts the wavelength of the monochromator, sets slit widths, and optionally changes flame conditions. The flame can even be switched between air/acetylene and nitrous oxide/acetylene within a run, although this requires a 15 s stabilization time. The system is completely computer controlled. The host computer is AT compatible with a 40 megabyte hard disk and 3.5 and 5.25 in. floppy drives. The computer has an EGA graphics display and 1 Mb memory. Instrument control is provided through the software, using a spreadsheet type of arrangement. This spreadsheet allows the operator to specify different parameters for each element, and to control the order in which they are analyzed and the order in which they are reported. There are also default values provided for each element at different concentration levels. An extensive wavelength catalog is included in the program to simplify the selection of alternative analytical lines. Sample results may be printed out or logged to the hard disk, or both.

Standards were prepared by diluting oil base standards (Conostan) one-to-four with kerosene (Fisher Scientific) by volume. Calibration was performed using four separate dilutions of Conostan 21 standards resulting in standards at 1, 5, 10, and 50 ppm levels. Ashless oil diluted with kerosene was used for the blank. Optimized mixtures of purified acetylene, AA grade (Airco), and nitrous oxide, USP (Airco), were used as the flame gases.

Results and Discussion

A hypothetical example is provided in order to compare analysis speeds of the various methods: the measurement of 20 samples for 20 elements. This analysis would take 100 min using conventional AA, not including time normally taken for lamp warmup. The analysis time is shortened by using sequential ICP to 84 min. Analysis by spark emission (rotrode) takes 51 min. Using the sequential AA under study, this analysis takes only 41 min. The calculation of the analysis times is presented in Tables 1 to 4 and is summarized in Table 5.

In these tables the measurement time per element is multiplied by 20. The result is indicated in the "Time/Sam" column. This is then multiplied by 20 again to indicate the total time for all samples. During the 2 s required to change the hollow cathode lamp, wavelength, and slit, the flame conditions can also be adjusted. The flame type can even be changed during the run. However, since 15 s are required to change flames, it is best to group elements so that this only needs to be done once per run. Figures 1 and 2 show that stability is obtained within 15 s of a flame switch, monitoring the copper absorbance during the switch.

TABLE 1—*Total analysis time (seconds) for conventional AA (analyses of 20 samples for 20 elements).*[a]

Scenario	Time/Element	×	Element	=	Time/Sample	×	Sample	=	Total Time
Sample intro.	5	×	20	=	100	×	20	=	2000
Sample changer	2	×	20	=	40	×	20	=	800
Determinations	4	×	20	=	80	×	20	=	1600
Recalibration	12	×	2	=	24	×	19	=	456
Changing element	NA				60	×	19	=	1140
Changing flame	NA				15	×	1	=	15
Total = 6011 s or 100 min									

[a]No time for lamp replacement was taken into account.

TABLE 2—*Total analysis time (seconds) for sequential ICP (analyses of 20 samples for 20 elements).*

Scenario	Time/Element	×	Element	=	Time/Sample	×	Sample	=	Total Time
Sample intro.	NA				30	×	20	=	600
Sample changer	NA				2	×	20	=	40
Determinations	10	×	20	=	200	×	20	=	4000
Recalibration	NA					NA		=	0
Changing element	1	×	20	=	20	×	20	=	400
Changing flame	NA					NA		=	0
Total = 5040 s or 84.0 min									

TABLE 3—*Total analysis time (seconds) for rotrode analyses of 20 samples for 20 elements.*

Scenario	Time/Element	×	Element	=	Time/Sample	×	Sample	=	Total Time
Sample intro.	NA				30	×	20	=	600
Sample pre-burn	NA				60	×	20	=	1200
Determinations	NA				10	×	20	=	200
Recalibration	NA					60		=	60
Changing electrodes					60	×	20	=	1200
Changing flame	NA					NA		=	0
Total = 3260 s or 54.3 min									

TABLE 4—*Total analysis time (seconds) for ANALYTE 16 (analyses of 20 samples for 20 elements).*

Scenario	Time/Element	×	Element	=	Time/Sample	×	Sample	=	Total Time
Sample intro.	NA				5	×	20	=	100
Sample changer	NA				2	×	20	=	40
Determinations	4	×	20	=	80	×	20	=	1600
Recalibration	NA					NA		=	0
Changing element	1	×	20	=	20	×	20	=	400
Changing flame	NA				15	×	20	=	300
Total = 2440 s or 40.6 min									

TABLE 5—*Summary of analyses time.*

	ANALYTE 16	Conventional AA	Sequential ICP	Rotrode
Seconds	2450	6011	5040	3260
Minutes	40.6	100.2	84.0	54.3
	1.0×	2.5×	2.0×	1.4×

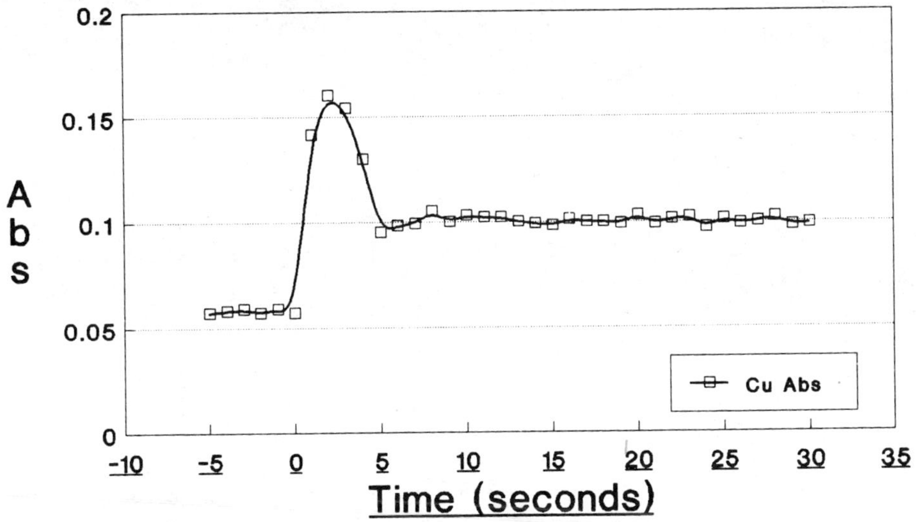

FIG. 1—*Flame equilibration (nitrous oxide to air).*

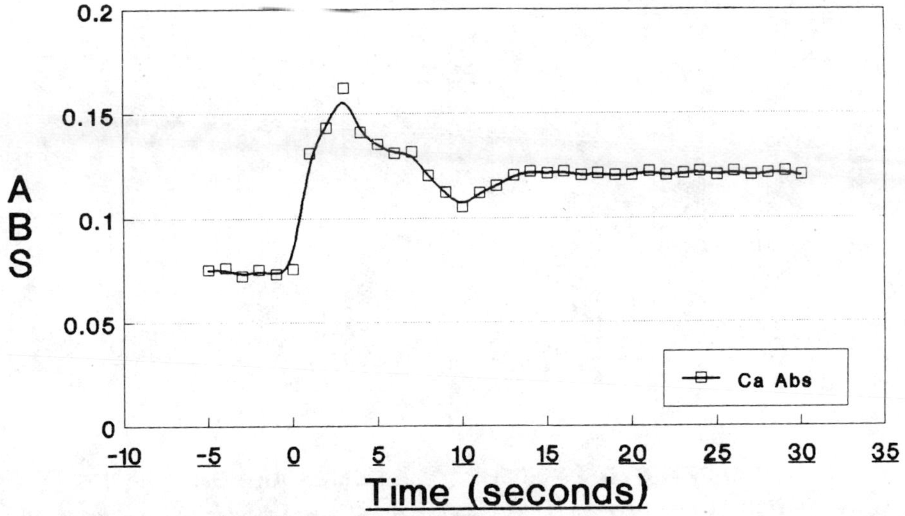

FIG. 2—*Flame equilibration (air to nitrous oxide).*

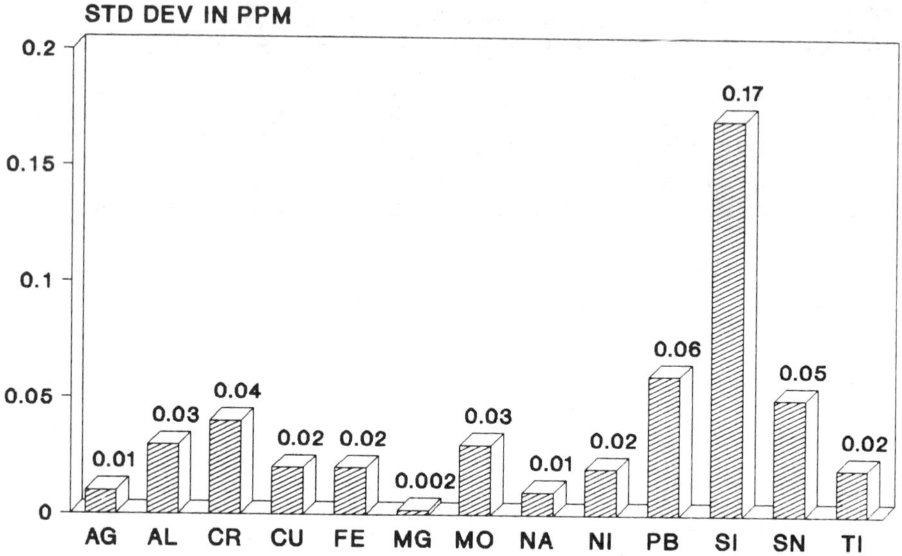

FIG. 3—*ANALYTE 16 multielement analysis. Detection limits for lube oil/kerosene; 10 runs on 5 ppm standard (single flame condition).*

Detection limits for ANALYTE 16 are shown in Fig. 3. These detection limits were obtained by taking 10 runs on a 5 ppm multielement lube oil standard. Sample was diluted with four parts of kerosene to one part sample. Detection limits were calculated as one standard deviation in ppm of the element in the diluted sample. Only the acetylene/nitrous oxide flame was used. Using the air/acetylene flame would have improved the detection limits for 8 of the 12 elements listed (all except Al, Mo, Si, and Ti). However, detection limits are not usually a concern in this analysis, so only one flame condition was used.

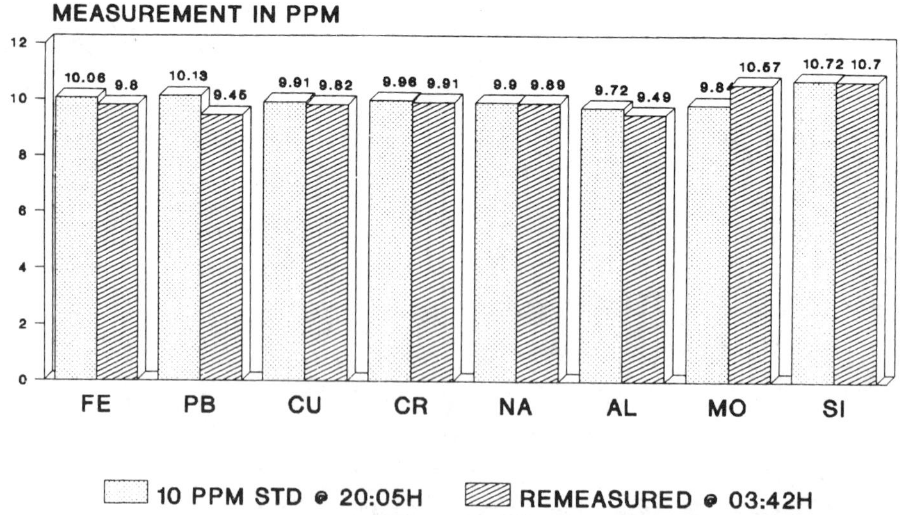

FIG. 4—*ANALYTE 16 multielement analysis: stability test (1 day of running on 1 calibration; 10 ppm Conostan standard/kerosene).*

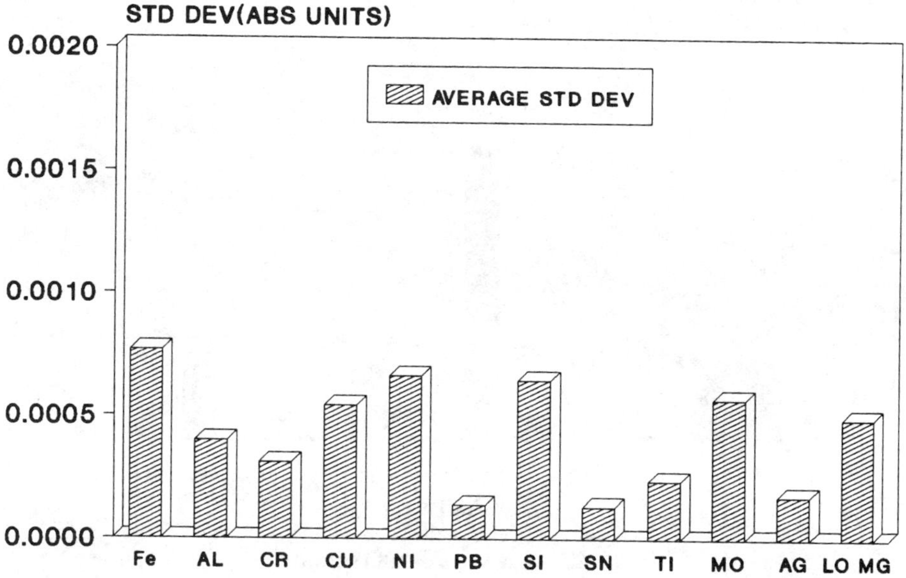

FIG. 5—*Conostan standards calibration: measurement precision (analyses of 60 jet oils).*

The precision and stability of the system was measured. The sequential mode requires that each lamp be kept in a warm state by using standby current. During the measurement phase, the lamp is pulsed rapidly. During the pulses, radio frequency (RF) energy is injected into the lamp. This results in very bright, narrow line sources. The end result of the bright narrow lines is good precision and stability. Figure 4 shows a stability test using a 10 ppm standard over an 8 h day without restandardization. A test of instrumental precision is shown

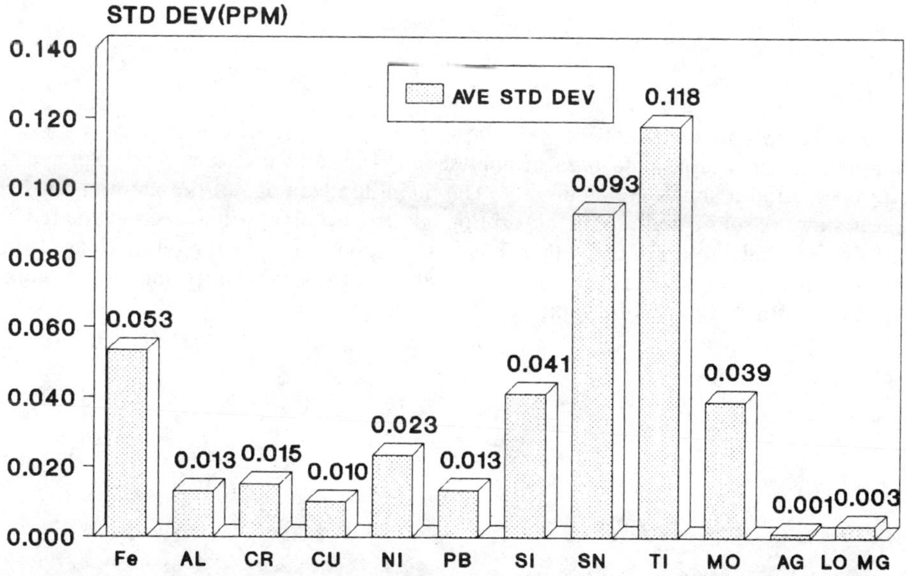

FIG. 6—*Conostan standards calibration: element precision (analyses of 60 jet oils).*

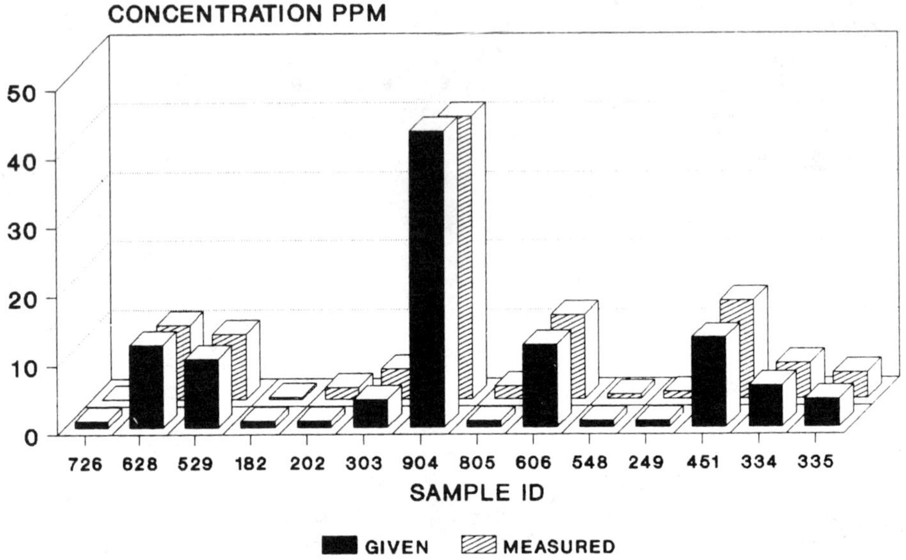

FIG. 7—*ANALYTE 16 multielement analysis of copper in light lube oil.*

in Fig. 5. Sixty jet oils were analyzed, and the average standard deviation for each element is shown in absorbance units. All are less than 0.001 absorbance units, which is the minimum reading on conventional AAs. The sequential AA tested has a measurement precision of a factor of ten better. The same data are shown in Fig. 6 expressed in parts per million of analyte. The bar identified as "LO MG" indicates that precision measurements were made only on those samples with low levels of Mg. Accuracy is illustrated in Fig. 7. A series of 14 oil samples was analyzed for copper. "Known" values are as reported by an industrial laboratory using a conventional AA.

Conclusion

Atomic absorption (AA) analysis of oil for wear metals has several advantages over other techniques. It has a large data base of analyses, it is easy to use, it is free from spectral interferences, and it shows great stability. The main disadvantage of the method has been that it is slow. A sequential atomic absorption spectrophotometer has been evaluated and found to retain all the advantages that make AA a good method for oil analysis. It also overcomes the speed handicap of conventional AA by significantly increasing sample throughput through the use of a rapid sequential analysis mode.

Danton Nygaard,[1] Francis Bulman,[1] and Timothy Alavosus[1]

Correlation Between Inductively Coupled Plasma and Rotating Disk Electrode Spark Emission Spectrometry for Used Diesel Engine Oils

REFERENCE: Nygaard, D., Bulman, F., and Alavosus, T., **"Correlation Between Inductively Coupled Plasma and Rotating Disk Electrode Spark Emission Spectrometry for Used Diesel Engine Oils,"** *Modern Instrumental Methods of Elemental Analysis of Petroleum Products and Lubricants,* R. A. Nadkarni, Ed., American Society for Testing and Materials, Philadelphia, 1991, pp. 77–82.

ABSTRACT: Fifty used and twenty new diesel engine oils are analyzed by both inductively coupled plasma and rotating disk electrode spark emission spectrometry. The results obtained by the two techniques are statistically correlated for ten wear metals and for four additive elements, and the implications of the correlations are discussed.

KEY WORDS: inductively coupled plasma emission, rotating disk electrode spark emission, used oil analysis, wear metals

Spectroscopic analysis of used oils for wear metals is an inexact science. Used oils are slurries containing suspended metallic particles, but spectrometers are calibrated using solutions of organo-metallic complexes dissolved in an oil base which is neither chemically nor viscosity matched to the samples. This matrix mismatch between samples and standards causes well recognized analytical errors [*1–3*]. However, despite inaccuracies, spectroscopic analysis of used oils is accepted as providing valuable information concerning unusual or excessive engine wear and is widely used to predict imminent engine failure [*4,5*]. Such conclusions concerning engine condition are drawn by comparing experimentally determined wear metal concentrations against a historical database for engines and oils of the same type. Unfortunately, reliance on a database for evaluation of data significance inhibits adoption of new methodologies which do not give results equivalent to, or at least correlated with, those obtained by an established method. This work investigates the relationship between analytical results obtained by a well established used oil-wear metals analysis technique, rotating disk electrode (rotrode) spark emission spectroscopy, and those produced by a relatively new technique, inductively coupled plasma (ICP) emission spectroscopy.

Development of rotrode emission spectroscopy for evaluation of wear metals in used oil began in the 1940s. It is a fast multi-element analysis technique that can be satisfactorily executed by relatively unskilled operators. Analytical results produced by the technique are precise, if not accurate, and it has found widespread use in preventive maintenance programs applied to diesel, gasoline, and turbojet engines, and to heavy machinery. However, the

[1] Baird Corporation, Bedford, MA 01730.

technique is labor intensive, requiring the continuous attention of a technician, and it does not easily lend itself to automation.

ICP emission is a comparably fast and precise multielement analysis technique. It is more sensitive than rotrode emission by several orders of magnitude, and it provides linear response over a wider range of concentrations. Although oil samples must be diluted with a low viscosity solvent (such as xylene or kerosene) prior to ICP analysis, both dilution and sample delivery can be automated. It is this attribute that makes ICP emission most attractive. However, it has been reported that ICP gives results different from those obtained by rotrode emission, and many long time rotrode users are reluctant to abandon their historical databases and begin anew with ICP.

Experimental Procedure

Samples

Fifty used oil samples and twenty unused oils were collected for analysis. The oils were all 10W40 weight, and the used oils were taken from diesel truck engines as part of a routine preventive maintenance program. The oils were analyzed neat by rotrode emission, but were diluted 1:5 by volume with kerosene prior to analysis by ICP emission. Ten wear metals (Fe, Cr, Cu, Mo, Al, Pb, Si, Ni, Sn, and Ag) and four additive elements (Mg, Ca, P, and Zn) were determined in each sample by both techniques.

Equipment

Rotrode analyses were performed on a Baird FAS-2C spectrometer, a commercial version of the rotrode spectrometer used by the U.S. military for used oil analysis. Analysis by ICP was performed on a Baird Plasma PST simultaneous spectrometer equipped with an IDS injection dilution system, an autosampler device that automatically dilutes as part of the sampling process [6]. Both instruments were calibrated using Conostan 75 base oil and S-21 mixed element standards prepared on a weight-weight basis at 10, 30, 50, 100, and 300 ppm for wear metals, and single element standards at 100, 300, 500, 1000, and 3000 ppm for additive elements.

Results

As shown in Table 1, all four additive elements were present in all used oils. Also, all used oils contained significant non-zero concentrations of iron. Other wear metals were present in only a fraction of the used oils, as shown in Table 2, and none of the used oils contained nickel, tin, or silver. Finally, none of the new oils contained any detectable wear metals by either analysis technique.

TABLE 1—*Summary of used oil results for additives.*

Element	Concentration Range, ppm[a]	No. of Non-Zero Readings
Mg	46–1326	50
Ca	1736–4136	50
P	1058–1417	50
Zn	1114–1926	50

[a]Determined by rotrode emission.

TABLE 2—*Summary of used oil results for wear metals.*

Element	Concentration Range, ppm[a]	No. of Non-Zero Readings
Fe	10–250	50
Cr	0– 29	8
Cu	0– 73	24
Mo	0– 19	11
Al	0– 21	10
Pb	0– 49	15
Si	0– 33	17
Ni	...	0
Sn	...	0
Ag	...	0

[a]Determined by rotrode emission.

Analytical results for all analyte elements, both wear metals and additives, were lower by ICP than by rotrode emission. A linear least-squares regression analysis applied to iron results for the fifty used and twenty new oils produced a correlation line with a slope of 0.70, an intercept of −0.38, and a correlation coefficient of 0.996. The remaining non-zero wear metal results are comparable to iron and are plotted on the iron correlation line in Fig. 1. Nickel, tin, and silver are omitted because they did not provide any non-zero results.

Additive elements did not respond in the same way. Instead, the ICP-to-rotrode correlation line produced by a linear least squares regression analysis has a slope of 0.88, an intercept of −3.6, and a correlation coefficient of 0.993. This line encompasses all four additive elements over the concentration ranges shown in Table 1.

Discussion

There is a clear correlation between ICP and rotrode results for both additive elements and wear metals, although the correlation relationship is distinctly different for the two groups. All four additive elements are well fit by a single correlation line, and all wear metals are well fit by a second correlation line. From these excellent correlations, it can be inferred that, within limitations, a database accumulated by rotrode emission spectroscopy can be applied to interpret analytical results obtained by ICP emission. As a first limitation, care must be taken to differentiate between wear metals and additives, because the two groups of elements respond differently. Secondly, because both analytical techniques are affected by sample viscosity, it is safe to assume that correlation lines relating the techniques are also viscosity dependent. For example, preliminary studies on heavy gear oils have given much different correlation slopes than those presented here for diesel engine oils. The internal standard method has sometimes been used in both ICP and rotrode emission spectroscopy to correct for viscosity induced error, and the equivalent response of all four additive elements strongly suggests that the method would be effective. However, an internal standard was not used in this work, because the extra time and labor involved causes the technique to be rarely used by high volume used oil laboratories. Therefore it would be prudent to apply the results of this study only to 10W40 diesel engine oils and to replicate the study for each additional oil type of interest.

It remains to be seen to what extent correlation between techniques is independent of minor changes in instrument operating conditions and therefore to what extent the correlation slope remains constant over time. It has been shown [2] that ICP response to particulates is highly dependent on instrument design considerations, such as nebulizer type

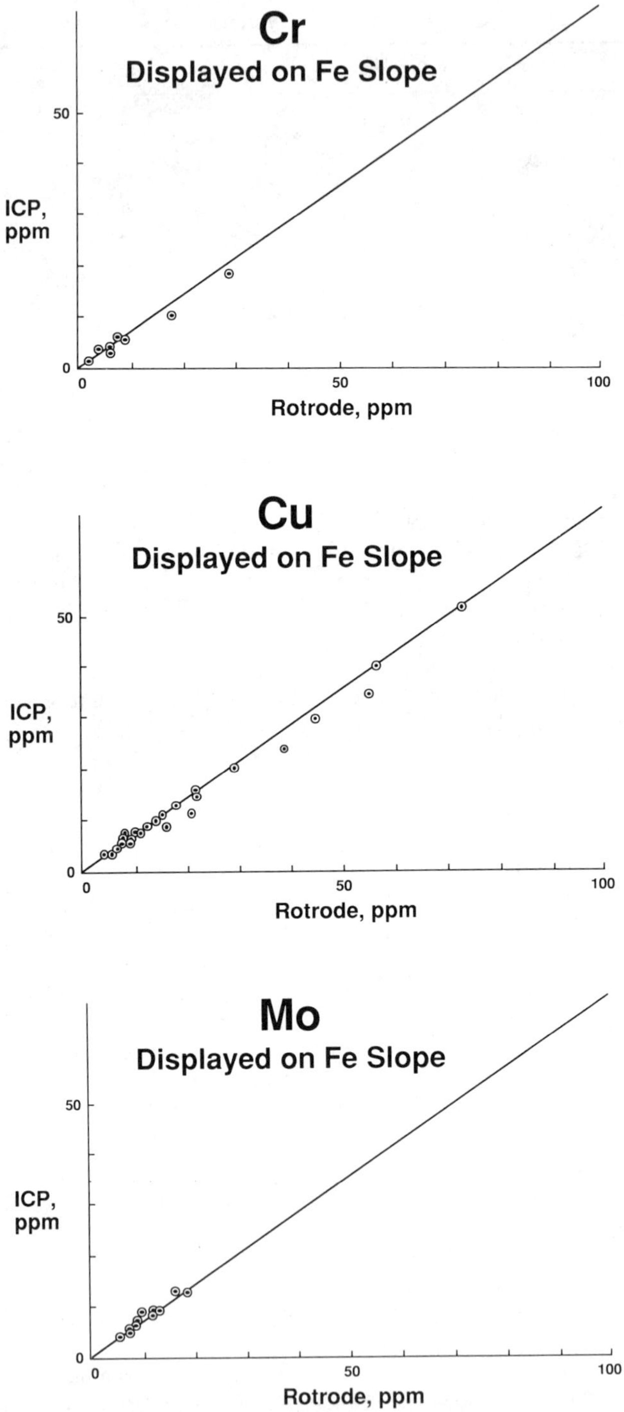

FIG. 1—*Non-zero wear metal results by rotrode emission plotted against ICP emission results and aluminum, lead, and silicon.*

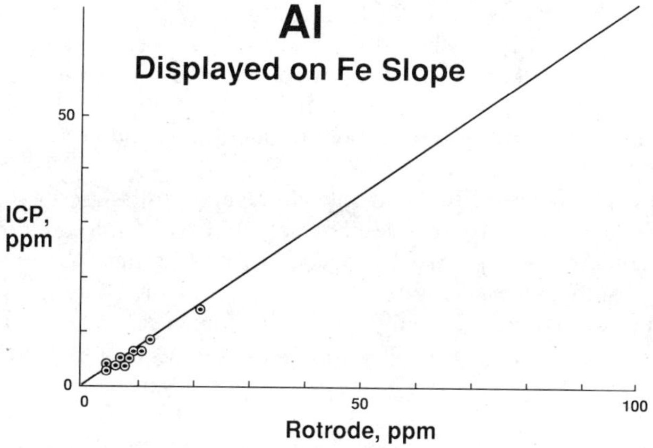

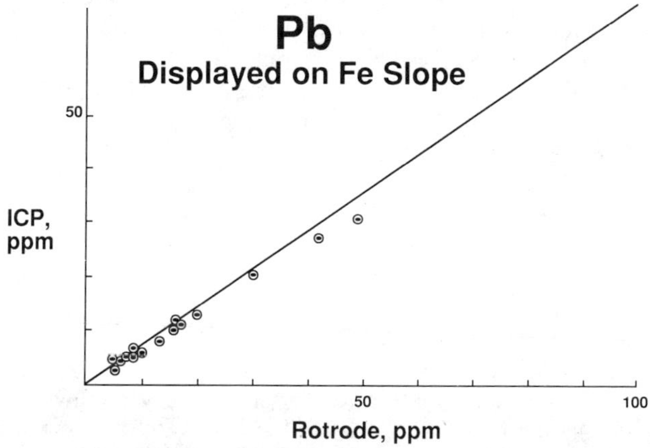

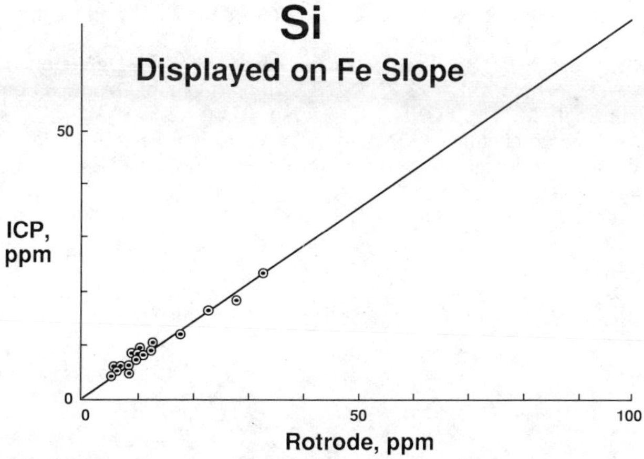

displayed against the iron linear regression correlation line for chromium, copper, molybdenum, and spray chamber configuration. However, the effects of ICP operating conditions, such as power and gas flows, on particulate response have not been studied. Independent ICP analyses by different operators of ten of the used oil samples chosen at random produced results that were in agreement with those reported here to within ±5%. This would indicate that the correlations are relatively independent of operating conditions, but more long term data are required to prove this point.

The non-equivalence between ICP and rotrode results can be ascribed to differences in the way that the two emission sources respond to (*a*) viscosity mismatch between samples and calibration standards, and (*b*) particles. The effect of particles is not significant to the additive element results, because additives are present in true solution as organo-metallic complexes. Therefore, for additives, only viscosity mismatch can be used to explain the non-equivalence between ICP and rotrode. Dilution with kerosene prior to ICP analysis tends to normalize sample and standard viscosities to that of the kerosene solvent; therefore ICP would be expected to be less sensitive to viscosity mismatch. Independent analysis of additives in another set of used oils samples by X-ray fluorescence confirmed the accuracy of ICP, so it can be concluded that the rotrode results for additives reported in this work are erroneously high (by approximately 10%) due to viscosity mismatch.

If it is assumed that matrix mismatch has the same effect on wear metals as on additives, then approximately one third of the 30% difference between ICP and rotrode for wear metals can be attributed to this cause. However, approximately two thirds of the difference between ICP and rotrode must be attributed to another cause, such as particulates. Size of particulates will affect both the efficiency of sample transport to the emission source and the ability of the source to completely atomize the sample. This work does not provide any way to differentiate between sample transport and atomization efficiency, but the results strongly suggest that rotrode is at least somewhat more effective in sampling particulates than is ICP emission.

References

[*1*] Rhine, W. E., Saba, C. S., and Kauffman, R. E., "Spectrometer Sensitivity Investigations on the Spectrometric Oil Analysis Program," Report No. AD-A 127-969/4/XPS, 1983 (available from National Technical Information Service, Springfield, Va., as NDN Order No. 033-0083-5089-5).

[*2*] Saba, C. S., Rhine, W. E., and Eisentrant, K. J., *Analytical Chemistry,* Vol. 53, 1981, pp. 1099–1103.

[*3*] Eisentrant, K. J., Newman, R. W., Saba, C. S., Kauffman, R. E., and Rhine, W. E., *Analytical Chemistry,* Vol. 56, 1984, pp. 1087A-1094A.

[*4*] Bond, A. and Kittinger, D. C., "Spectrometric Oil Analysis for the Detection of Incipient Turbojet Engine Failures," Report No. AD-472 771/5/XPS, 1965 (available from National Technical Information Service, Springfield, Va., as NDN Order No. 033-0042-3964-3).

[*5*] O'Sullivan, D. F., "Spectrochemical Oil Analysis," Report No. PB-216 168/5/XPS, 1973 (available from National Technical Information Service, Springfield, Va., as NDN Order No. 033-0014-8999-6).

[*6*] Evans, S. J. and Klueppel, R. J., *Spectrochemica Acta,* Part B, Vol. 40B, 1985, pp. 49–55.

Malte Lukas[1] *and Daniel P. Anderson*[1]

Techniques to Improve the Ability of Spectroscopy to Detect Large Wear Particles in Lubricating Oils

REFERENCE: Lukas, M. and Anderson, D. P., **"Techniques to Improve the Ability of Spectroscopy to Detect Large Wear Particles in Lubricating Oils,"** *Modern Instrumental Methods of Elemental Analysis of Petroleum Products and Lubricants, ASTM STP 1109,* R. A. Nadkarni, Ed., American Society for Testing and Materials, Philadelphia, 1991, pp. 83–95.

ABSTRACT: Machine condition monitoring through oil analysis is a proven and cost-effective maintenance management technique. Periodically, however, the ability and need of the analytical techniques to detect large wear particles are questioned. Today there is general consensus as to the maximum wear particle size capability of atomic emission spectrometers (AES) and atomic absorption spectrometers (AAS), but few agree on the ways to improve them. This paper discusses three techniques that can be used to improve the ability of optical emission spectrometers to detect and quantify large wear particles in lubricating oil.

The *single spark technique* is based on an atomic emission spectrometer readout system that stores the intensity of individual excitation sparks for each element and presents the data as a particle size distribution. The system can operate in real time and provide information on each analysis, or the output can be suppressed and the user alerted only if wear particles exceed a preset size limit. Test data has shown an ability to detect and identify wear particles up to 40 μm in size.

The *ashing rotrode technique* is based on a sample introduction and analysis system in which the oil sample is deposited on a graphite electrode, then resistively heated to desolvate the oil matrix. The desolvated matrix is subsequently excited by a pulsed D-C arc, and the spectra is analyzed by a polychromator. This technique minimizes matrix effects, enhances spectral sensitivity, and greatly improves detection of large wear particles.

The *acid digestion differential technique* is based on two analyses of the same oil sample. The sample is analyzed with and without acid to dissolve wear particles. The difference in the two readings indicates the presence of large wear particles.

All three techniques discussed in this paper improve the ability of spectrometric techniques to detect, quantify, and identify wear particles. Capabilities and limitations of each are discussed. The technique of choice varies with the requirements and philosophy of the analytical laboratory.

KEY WORDS: oil analysis, wear metal particles, ICP, inductively coupled plasma, atomic emission spectrometer, AES, atomic absorption spectrometer, AAS, rotating disc excitation, RDE, single spark, ashing rotrode, acid digestion

Many excellent reports and articles have been written on the subject of oil analysis and its historic evolvement. Simply stated, oil analysis is a diagnostic maintenance tool used to detect and quantify wear metals and contaminants in the lubricant of oil wetted systems. It is based on the fact that a representative sample of lubricant, when periodically analyzed, provides important information on the condition of the wearing surfaces. It is a test that

[1] Spectro Incorporated, Littleton, MA 01460.

helps to determine whether a mechanical system is in a normal wear mode, or whether a potentially serious problem exists.

Wear particles are generated as a result of friction and the continuous wearing away of contacting surfaces. They enter the lubricant in the form of metallic particles from submicrometer to millimeter in size. Although present day oil analysis enjoys a high rate of success, existing emission spectrometric techniques are blind to large wear particles [*1,2*]. The particle size limitation varies with the analytical technique, but general agreement places it at particle sizes in the range of 5 to 10 μm [*2,3*].

The particle size limitation of spectrometric techniques was first identified and openly discussed in great detail at an oil analysis symposium in 1978 [*1,2*]. Since that time, several studies have been done on the subject. Unfortunately, this phenomenon suffers from inadequate public relations, and the analytical size limitation of atomic emission spectrometers seems to have been rediscovered about every two years since then.

This paper presents several methods that have been developed over the last ten years to improve the large particle detection capabilities of emission spectrometers. The ultimate goal is to detect the relatively small percentage of failures missed by routine spectrometric oil analysis techniques.

Influencing Factors

Detection of large particles by atomic emission spectroscopy is influenced by a number of factors [*4*]. The most important are described below:

1. *Representative Sample*—If a system is in a severe wear mode and generating large wear particles, the oil sample taken to the laboratory for analysis must contain a representative quantity and size distribution of particles. The size of the sample and good sample taking technique can be an important factor. Likewise, the sample must be vigorously shaken to resuspend particles before measurements are made in the laboratory.

2. *Excitation Parameters*—The type of excitation influences the particle size detection capability of spectrometric techniques. Excitation power to dissociate a particle is the key factor. However, the hottest source, an inductively coupled plasma (ICP), is limited by short sample residence time in the plasma. Depending on composition, particles may not be in the plasma long enough to dissociate and vaporize. For the rotating disk electrode (RDE) method, an arc source has better efficiency for certain elements, but for others a spark source is preferred.

3. *Oil Viscosity*—For the RDE method, the viscosity of the oil has an effect on the quantity of the sample reaching the excitation process and thus the quantity of wear particles carried with it. This is further influenced by the porosity of the disk electrode. Viscosity is less of a factor with AAS or ICP-AES, since most oil samples must be diluted before analysis, eliminating viscosity as a major variable.

4. *Morphology of Wear Particles*—The shape of a particle has an effect on the efficiency of the spectrometric technique. For example, rubbing wear particles have a length-to-thickness ratio of about 10:1. Powdered metals are often more-or-less spherical. If both types are described by their major dimension, spherical particles of the same dimension as wear particles will be more difficult to excite.

5. *Lubricant Composition*—It has been shown that the efficiency of the analytical technique varies between hydrocarbon oils and ester oils [*1*].

6. *Quantity of Sample Introduced into the Excitation Process*—The ability to measure large wear particles presumes that they are present in the small quantity of sample that is actually excited. For a typical 1 mL oil sample, 100% of it is subjected to analysis with the

ashing technique (discussed in this paper), less than 50% for the RDE method, and only a few percent for the ICP method. If there are only a few large particles present per milliliter of sample, the probability of detecting those particles is thus greatest for the ashing technique.

7. *Method of Sample Introduction*—In the ICP method, the sample is nebulized to form an aerosol. By the time the aerosol reaches the plasma, the droplet size is very fine, typically less than 5 μm. Consequently, wear particles larger than 5 μm have no chance of being detected.

8. *Standard Samples for Particle Size Investigations*—A recurring problem in experimental work with wear particles is obtaining oil samples which contain wear particles of defined sizes. In some of the work described below, powdered metals were purchased, then separated according to size with a sonic sifter and nuclepore filters at 5, 10, and 20 μm to obtain separate filtrates with known maximum particle size. These powders were subsequently mixed, by weight, with MIL-L-7808 or MIL-L-23699 synthetic lubricant. The particle size distributions were then verified using a particle counter and by scanning electron microscopy.

Any new method to improve the large wear particle detection capabilities and efficiency of the AES techniques must address many of the above influencing factors. Several such methods are described below.

Single Spark Technique

Description

The single spark technique is not a new development; it is a new application for an existing spectrometer data readout system [5]. It is based on the principle that if metallic particles are present in a fluid and are excited with a single arc or a spark, the intensity of light emission will be proportional to the size of the particle.

A single spark system can be used with almost any spectrometer. It takes the output from the photomultiplier tubes and utilizes fast integrators to store and quantify the intensities of up to 2000 individual sparks for each element during an analysis. In a production environment, the actual data from such a system are not displayed during the analysis cycle. The spectrometer provides normal data readout in parts per million, and single spark data are used to provide an alarm or other type of indication if the intensity (particle size) for an element exceeded an operator adjustable threshold.

Raw data for a single spark system are in the form of an intensity distribution. Increasing intensity corresponds to larger particle size. Readout is available on a CRT, or on a printer for hard copy. A typical readout for one element is shown in Fig. 1. The abscissa shows, from left to right, increasing intensity of individual sparks and corresponds to particle size. The scale can be calibrated to actual particle size. The ordinate corresponds to the number of sparks at each intensity. It is therefore a plot of particle size versus the number of particles at each size.

The single spark software has a menu to allow the user to select from a number of statistical variables such as total sparks, mean, median, integral, maximum, etc. Excitation source parameters can also be varied by the software to select either 60, 120, or 240 sparks per second. For the experimental work presented in this paper, 120 sparks per second was used with a total integration of 1000 individual sparks. Therefore, analysis time for each sample was less than 9 s.

Analytical Results

Initial tests were performed with a conventional rotating disk sample introduction system. Oil samples, prepared to contain the various particle sizes, were analyzed and very little, if

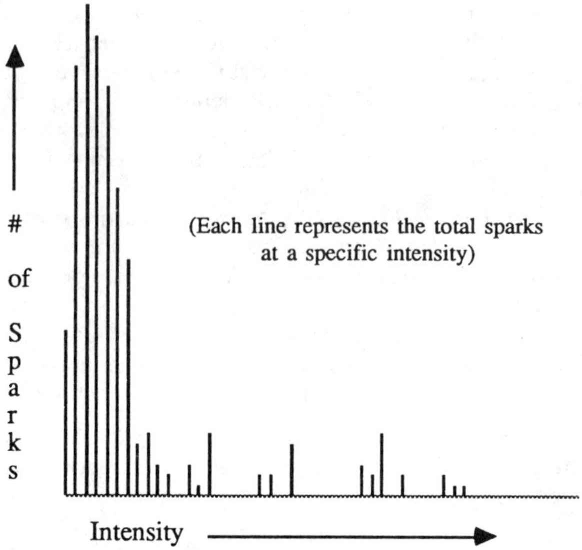

FIG. 1—*CRT readout for single spark method.*

any, difference was noted in the intensity readouts for samples with particles greater than 10 μm. These results reconfirmed one of the wear particle related shortcomings of the RDE technique, namely that there are large particle transport difficulties.

The single spark system is ineffective without an efficient sample introduction system. Experimental modifications to the sample introduction system were an immediate focal point of additional work. To allow more of the metallic particles to enter the area of excitation, several techniques were tried including hollow counter electrodes to direct the sample onto the disk electrode, scalloped disks to "shovel" more of the sample into the excitation, and disks with varying porosity. These experiments led to a very simple and practical solution. Square electrodes were manufactured with a "trough" to contain and hold, for each analysis, the same quantity of oil sample. This solved the problem of carrying all the sample into the arc. Furthermore, after a few minutes the porous graphite electrode absorbed the oil, leaving only the analyte.

In order to obtain a base level for the analytical data, the single spark signature for MIL-L-7808 without any metal particles was determined. Figure 2 shows two single spark readouts,

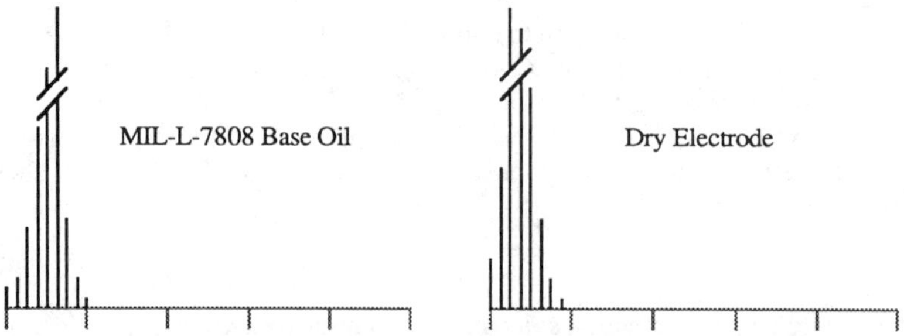

FIG. 2—*Single spark signature for base oil and dry electrodes.*

one with MIL-L-7808 oil and the other of the electrodes alone, without any oil. The intensities closest to the origin were determined to be background signal. The background is similar for most elements and is attributable mostly to the electrodes themselves.

With an understanding of the background signal, the readout for oil samples with actual metallic particles is easier to understand. Figure 3 is a single spark signature for the analysis of 100 ppm chromium oil samples with metallic particles at <10, <20, and <74 μm in size. Clearly there is an increase in the intensity distribution that corresponds to particle size. The background signal for chromium (Fig. 2) has all spark intensities within the first two decades. The sample with particles below 10 μm has spark intensities which extend well into the third and fourth decades. The 20 μm sample contains spark intensities over five decades. Finally the sample with particles up to 74 μm covers all decades with spark intensities.

All three samples were prepared to contain, by weight, 100 ppm of chromium. One would therefore expect that the integral of the intensities, which corresponds to normal ppm readout, to be the same. Although the integral was within experimental error for the samples with the smaller particles, it was only half as much for the sample with the particle distribution of up to 74 μm. Repeated analyses produced similar results. It is therefore probable that, although some of the larger particles are excited by the spark and provide an intensity readout, many of the particles are too large to be completely volitalized.

The experiments were repeated for other elements and similar single spark signatures depicting large particles were obtained, thus demonstrating that with proper sample introduction the single spark technique works well in identifying large particles which are otherwise not detected.

A similar system, based on the same principle, was discussed by Kaufman [*6*] in the development of a multielement data acquisition system. He demonstrated that by monitoring voltage buildups on integrating capacitors, the spectrometer could differentiate between dissolved wear species produced by normal wear modes and wear debris produced by abnormal wear modes. Qualitative particle size distributions and particle composition analyses were performed on metal powders prepared in MIL-L-23699 lubricating oil. As in the work described above, the efficiency of the spark source to fully vaporize large wear particles was also questioned. Nevertheless, the system was shown to have the potential to improve abnormal wear detection capabilities of oil analysis programs.

Ashing Rotrode Technique

Another method, the ashing rotrode technique, was also investigated towards improving the particle size detection capabilities of emission spectrometers. In this technique, the oil

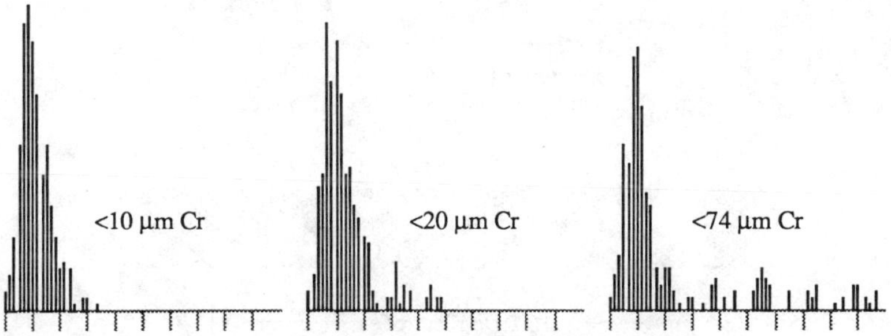

FIG. 3—*Single spark signature for chromium particles in oil at various size levels.*

sample is deposited directly on a graphite electrode, then resistively heated to desolvate the oil matrix. The desolvated ash is then excited by a pulsed d-c arc. The resultant spectra is analyzed by simultaneous AES. Preliminary experimental data show that the ashing rotrode technique improves detection of large wear particles, minimizes matrix effects, and enhances spectral sensitivity.

The ashing rotrode technique was based on earlier work which had shown that manipulations to excitation source parameters and sample introduction systems will improve particle detection capabilities of AES. It had been determined that true improvements in detection capabilities were possible with slower disk electrode speeds, higher excitation potential, and by excitation of samples after the oil was absorbed into the pores of the electrode [*4,5*]. In developing a particle detection system, it was decided that the first step was to desolvate the oil matrix, followed by exposure of the analyte to an excitation plasma suitable for efficient vaporization of the elements of interest.

Preliminary work on this method showed that it was possible to quantitatively analyze wear particles up to 40 μm in size by manual and time-consuming ashing techniques [*7*]. Through desolvation and concentration of analyte, the ashing rotrode technique could produce a limit of detection for most elements at the sub part per million level. This concept, coupled with further automation, led to the ashing rotrode system.

Description

Preliminary research into the development of the ashing rotrode technique revealed that variations in specific density of individual graphite electrodes would produce inconsistent ash residue [*8*]. To minimize this effect, 12.5 mm diameter graphite stock was cut to lengths of 100 mm and machined such that 10 individual electrode surfaces of 5 mm in width were spaced 2.5 mm apart. This became the ten sample ashing rotrode (Fig. 4). The rotrode was then mounted horizontally between two chucks in the spectrometer's sample excitation chamber. The chucks were driven by a synchronous motor which rotates the electrodes and provides electrical contact to an ashing source.

The second graphite electrode is 6 mm in diameter and acts as the counter electrode and anode. It is mounted on a slide mechanism which moves parallel to the ashing rotrode. This permits the anode to be positioned over any of the ten electrode surfaces for sequential analysis of up to ten separate oil samples. An optical quartz fiber mounted on the slide mechanism follows the movement of the counter electrode and directs the light from the excitation process into a polychromator.

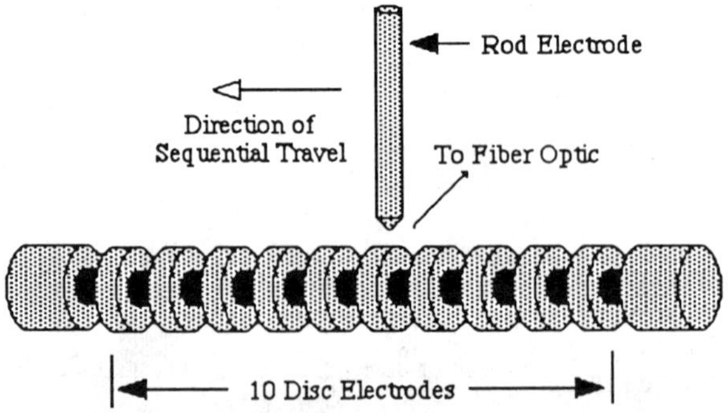

FIG. 4—*Ten sample ashing rotrode and counter electrode.*

The procedure to ash and analyze an oil sample begins while the electrode is rotating with the deposition of 30 μL of oil with a Teflon-tipped positive displacement pipet on the first lobe. When all nine samples have been applied, the ashing current is automatically initiated and timing is controlled by a built in microprocessor. Typical ashing time for ten samples is 60 s. The analytical cycle is the same as in a normal RDE-AES and can be performed sequentially for each lobe, and immediately after the ashing step. Analysis of each lobe takes 30 s, during which time the electrode rotates one complete revolution.

Analytical Performance

It was apparent from the outset that proper sample deposition techniques had to be developed in order to obtain good experimental data. The performance obtained with the ashing rotrode technique was excellent and can be summarized as follows:

1. *Detection Limit*—Initial data produced excellent detection limits. Table 1 presents the detection limits of the conventional rotating disk technique based on 2σ compared with the ashing rotrode technique at 3σ. The detection limits appeared to be a function of the quantity of wear particles on the electrode lobe, since it was possible to improve and vary them by adding more of the oil.
2. *Matrix Effect*—The ashing rotrode technique eliminates the matrix effect cause by varying types of lubricants. A background correction system is therefore not required. Table 2 compares the background emissions of three different oil matrices.
3. *Accuracy*—The accuracy of the ashing rotrode on metallo-organic standards can be stated in general terms for most elements as 0.5 ppm, or 5% of the concentration, whichever is greater. The ashing rotrode, however, was not developed to analyze oil standards. It was developed to provide quantitative analysis of oil samples containing wear particles not readily detectable by existing instrumentation. For this reason, it is possible to provide analytical results on used oil samples not always consistent with other techniques. The technique was used as part of a study for the DoD Joint Oil Analysis Program, where it displayed good correlation to standard samples and those with additive concentrations in submicron components. In used oil samples containing particles, the ashing rotrode always predicted higher concentrations than existing analytical techniques. This is to be expected if the technique is more efficient in detecting wear particles, but it is not entirely clear if all the differences are due to actual wear particles alone.

TABLE 1—*Comparison of detection limits of standard rotrode and the ashing rotrode.*

Element	Wavelength, nm	Standard Rotrode, ppm	Ashing Rotrode, ppm
Iron	259.9	0.21	0.06
Silver	328.1	0.02	0.03
Aluminum	396.1	0.07	0.20
Chromium	425.4	0.36	0.30
Copper	324.7	0.09	0.03
Magnesium	280.2	0.01	0.01
Nickel	341.4	0.66	0.03
Silicon	251.6	0.38	0.80
Tin	317.5	0.88	0.20
Titanium	334.9	0.25	0.10
Molybdenum	281.6	0.79	0.15

TABLE 2—*Comparison of the effect of oil matrices on net emission.*

Element	MIL-L-7808H	Conostan 75 Base Oil	MIL-L-23699
Iron	0.9630	1.0000	0.9907
Silver	0.9756	1.0000	1.1168
Aluminum	0.8694	1.0000	0.8703
Chromium	0.9079	1.0000	0.8842
Copper	0.9266	1.0000	0.9174
Magnesium	1.0288	1.0000	1.0096
Nickel	0.9382	1.0000	0.9500
Tin	0.8654	1.0000	0.9317
Titanium	1.0826	1.0000	1.0826
Mean	0.9385	1.0000	0.9667

The ashing rotrode, along with other instruments and techniques, was used to analyze actual oil samples, many with wear particles, as part of a study for the DoD Joint Oil Analysis Program. Tables 3 and 4 provide an example of analytical results produced by the acid dilution method using inductively coupled plasma atomic emission spectroscopy (ADM ICP/AES), direct dilution method (DDM ICP/AES), the rotating disk electrode (RDE/AES), and the ashing rotating disk electrode (ARDE/AES).

Table 3 summarizes the analyses for the iron wear metals by all techniques in the study. The sample numbers with two digits are for oil samples from aircraft gas turbines and those with four digits are oil samples from diesel engines. The ashing rotrode provides fairly good correlation to the other techniques except for Samples 18 and 1493. For these two samples the acid dissolution method indicates slightly higher concentrations than the direct dilution

TABLE 3—*Comparison of iron analyses produced by four different analytical techniques.*

Sample No.	ADM ICP/AES	DDM ICP/AES	RDE AES	ARDE AES
1	1.21	0.80	1.1	1.40
4	4.40	3.70	3.9	3.86
6	5.64	4.7	5.0	7.36
8	4.32	3.67	3.8	4.62
12	6.01	4.58	4.3	11.4
13	2.00	1.53	1.3	3.59
17	2.21	1.49	1.5	5.38
18	4.10	2.83	3.4	10.20
21	0.93	0.55	2.0	1.11
28	2.95	2.30	4.0	3.14
29	4.20	3.51	1.0	5.02
1440	54.90	52.50	49.0	56.4
1442	42.60	40.00	35.0	48.70
1443	51.70	47.10	46.0	61.90
1491	57.50	62.00	49.0	61.20
1492	121.00	101.00	83.0	91.00
1493	39.90	33.90	27.0	51.4
1498	261.00	323.00	220.0	186.00
1501	63.50	59.10	49.0	81.60
1502	67.60	67.60	67.0	78.90

TABLE 4—*Comparison of chromium analyses produced by four different analytical techniques.*

Sample No.	ADM ICP/AES	DDM ICP/AES	RDE AES	ARDE AES
1	0.09	0.18	1.0	0.48
4	0.64	0.91	1.0	0.09
6	0.92	1.14	2.0	1.25
8	1.67	1.29	2.0	0.19
12	0.72	0.58	1.0	1.35
13	0.51	0.36	1.0	1.35
17	0.59	0.38	1.0	0.92
18	0.76	0.61	1.0	2.17
21	0.24	0.20	...	0.19
28	0.09	0.18	1.0	0.89
1440	3.27	3.10	1.0	5.11
1442	4.43	4.10	1.0	11.40
1443	10.00	10.20	7.0	26.50
1491	8.70	8.64	6.0	22.70
1492	17.30	19.70	14.0	33.50
1493	3.99	3.99	2.0	13.2
1498	15.50	16.80	12.0	44.60
1501	5.60	5.69	3.0	18.40
1502	6.50	6.88	4.0	17.00
1506	2.37	2.57	1.0	2.37
1507	2.00	2.00	1.0	3.11

and rotating disk methods; however, the ashing rotrode records significantly higher concentrations than both of these techniques. From all indications and a visual inspection, this is a situation where large wear particles exist in these samples and the routine techniques failed to detect them. This phenomenon is supported by the analytical data for chromium (Table 4). The data for the relatively clean oil samples from the aircraft turbines correlate very well between the various techniques. The analytical data for the diesel engine samples that contain large amounts of wear particles are, as expected, consistently higher for the analyses using the ashing rotrode technique. The exceptions are Samples 1506 and 1507. These two oil samples were clear in appearance and were new oils or could not have been in a system for any extended period of time to generate wear particles.

The analytical data generated by the ashing rotrode technique has stirred a great deal of interest and debate as to how the data should be interpreted. Nevertheless, it has been proven to be an effective method to improve the large wear particle detection capability of analytical instruments. Refinements in sample application, rotrode mounting, and the ashing control cycle are all areas for further investigation and improvement before the technique is placed into routine service in an oil analysis laboratory.

Acid Digestion Differential Method

It was recognized more than ten years ago that in spite of the extensive application of spectrometers for oil analysis, no analytical procedure was available which satisfied all requirements in regard to speed of analysis, precision, accuracy, sensitivity, convenience of sample preparation, and cost per analysis. Inductively coupled plasma (ICP) spectrometry was shown by Fassel and Petersen [9] to be more capable in satisfying all of these requirements. However, the ICP method, as routinely applied, has very poor particle sensitivity. An oil sample, diluted by solvent to lower viscosity, is nebulized to create an aerosol which flows through a spray chamber before reaching the plasma. The spray chamber removes all but the finest droplets by settling and impaction, so that the maximum droplet size reaching the plasma is approximately 5 μm. Consequently, particles in the sample larger than 5 μm have no chance of being detected.

Justification

If spectrometric measurements are made as a function of time after an engine oil change, the wear metals concentration will continue to increase indefinitely (Fig. 5) so long as clean make-up oil is not added to the engine. In fact, for spectrometric measurements to be meaningful, the time since the last oil change must be known, because it is the increase in wear metals concentration during a given operating period that is of significance. Spectrometric data for aircraft gas turbine engines are a special case. Absolute maximum concentration levels apply to gas turbine engines because even though the oil in most gas turbine engines is not changed, the wear metals concentration reaches a maximum due to the frequent addition of make-up oil. Kjer [*10*] describes the mathematics of wear particle concentration for small and large particles and for topping-off.

From the foregoing discussion, it may be seen that the total metals concentration in a lubricating oil sample is the total of two components: the fine particle and chemically dissolved portion and the large particle portion. The concentration of the fine particle and chemically dissolved fraction continues to increase between oil changes, but the large particle fraction reaches a dynamic equilibrium concentration, wherein the contribution of generated wear particles is offset by the loss of large particles to filters, settling, or comminution. The concentration of large particles is of the greatest interest in regard to abnormal wear, because the quantity and size of large particles increases substantially for many, and probably most, abnormal wear modes. Also, the contribution of the large particles to the total wear metals concentration is often a much smaller fraction than the contribution of the fine particles and dissolved material [*11*].

In order for spectrometric analysis to be as effective as possible in providing early indication of abnormal wear, it is not always sufficient to monitor only the total wear metals concentration by a particle size independent method. Anderson [*12*] suggests that a differential method be considered, at least for critical equipment having a history of rapidly developing failure modes, wherein two measurements are made on each sample. An ICP spectrometer is recommended for this procedure because of its superior precision, accuracy, and limits of detection. One measurement is made by directly introducing the sample to the plasma

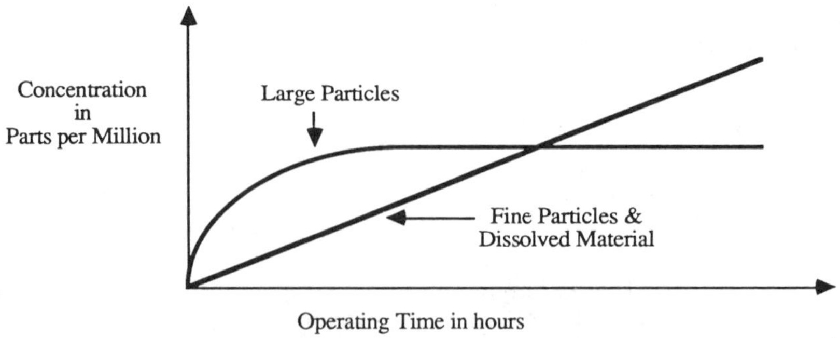

FIG. 5—*Particle concentration as a function of time.*

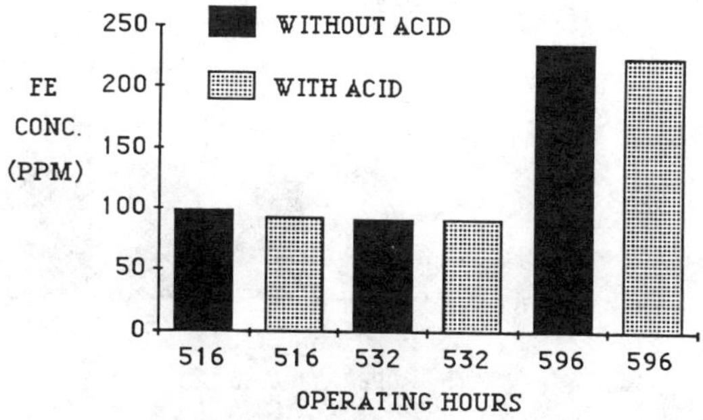

FIG. 6—*Iron results for Series A.*

after diluting it with solvent by an appropriate ratio. This measurement gives the fine particle fraction. The second measurement is made using a particle size independent method to arrive at a total metals concentration. The difference between the two measurements is an indication of large particle fraction.

The large particle fraction is expected to remain within certain limits during trouble free operation, even as the lubricant experiences more operating time. Abnormal wear modes, filter clogging, contaminant ingestion, lubricant failure, etc., are expected to significantly increase the large particle fraction.

Analytical Results

Two series of oil samples were obtained from reciprocating engine test stands. Series A samples show no increase in the severity of wear during the course of the test as determined by ferrographic analysis. The quantity of large particles is approximately the same in all Series A samples. Ferrograms prepared for Series B samples show a substantial increase in the concentration of large particles as the test proceeds.

Each sample from Series A and Series B was heated, homogenized by shaking, and split into two portions. One portion was left as is; an acid digestion following the method of Brown et al. [*13*] was performed on the other portion. All portions were weighed on an electronic balance, as was the acid and solvent (kerosene) added to each portion, so that concentration data could be accurately calculated. Each portion was then analyzed by ICP-AES so that two measurements were obtained for each sample: (1) ICP-AES without acid digestion, and (2) ICP-AES with acid digestion.

Figure 6 shows iron results for Series A samples. Little difference is obtained between

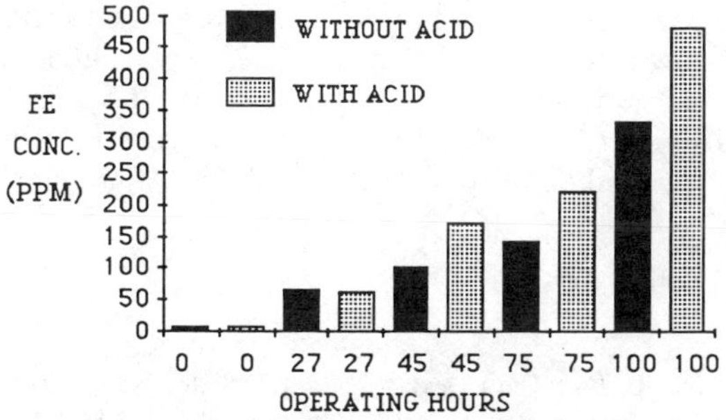

FIG. 7—*Iron results for Series B.*

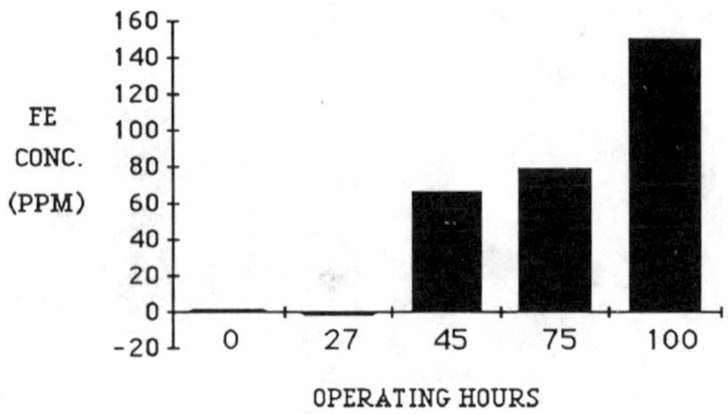

FIG. 8—*Difference between digested and undigested samples for Series B.*

measurements with and without acid digestion. Figure 7 shows progressively greater difference between the two methods for Series B samples. Figure 8 plots the difference between the two methods as the engine test proceeds.

The power of this differential acid digestion technique is that it is able to indicate the presence of large particles in situations where the concentration is increasing, as measured either by routine techniques or particle size independent techniques. In one case, Series A, there is no increase in large particles, only normal wear may be said to be on-going, whereas in Series B a progressive increase in wear severity is occurring. Without differential measurements, either case appears to show an increasing trend.

Conclusion

It has been shown that:

1. The single spark technique is very effective in detecting and quantifying large wear particles in oil samples. It will work with a standard RDE-AES spectrometer; to be effective, however, the sample introduction system has to be modified.
2. The ashing rotrode shows the most promise as a separate analytical technique. It can be incorporated into a modified RDE-AES spectrometer, but additional laboratory work is required to fully understand its capabilities.
3. The acid differential technique can identify, by element, the presence of large wear particles in lubricating oil samples. It is a two-step method, because the particle size independent technique by itself does not necessarily provide the needed information about abnormal wear.

Although the wear particle size limitation of atomic emission techniques is periodically "rediscovered," it is now safe to say that methods exist which circumvent this so-called limitation. This paper has shown three methods that improve the particle detection capability to varying degrees. The overall efficiency of oil analysis can be improved if any of these methods are applied. The method of choice varies with the requirements and the philosophy of the laboratory.

References

[1] Lukas, M. and Giering, L. P., "The Effects of Metal Particle Size in the Analysis for Wear Metals Using the Rotating Disc Atomic Emission Technique," presented at the International Symposium on Oil Analysis, Erding, Germany, July 1978.

[*2*] Eisentraut, K. J., Thornton, T. J., Rhine, W. E., Costandy, S. B., Brown, J. R., and Fair, P. S., "Comparison of the Analysis Capability of Plasma Source Spectrometers vs. Rotating Disc Atomic Emission and Atomic Absorption Spectrometry for Wear Particles in Oil: Effect of Wear Particle Size," presented at the International Symposium on Oil Analysis, Erding, Germany, July 1978.

[*3*] Rhine, W. E., Saba, C. S., and Kaufman, R. E., "Metal Particle Detection Capabilities of Rotating-Disc Emission Spectrometers," *Lubrication Engineering,* Vol. 42, No. 12, 1986, p. 755.

[*4*] Rhine, W. E., Saba, C. S., and Kaufman, R. E., "Spectrometer Sensitivity Investigations of the Spectrometric Oil Analysis Program," NAEC Report 92-169 Naval Air Engineering Center, 22 April 1983.

[*5*] Lukas, M., "Development of a Particle Size Independent Method for Wear Metal Analysis," presented to the International Oil Analysis Symposium, Department of Defense, Pensacola, Fla., May 1983.

[*6*] Kaufman, R. E., "Particle Size and Composition Analyses of Wear Debris Using Atomic Emission Spectrometry," *Lubrication Engineering,* Vol. 45, No. 3, 1989, p. 147.

[*7*] Gilbert, T. R., Sondoval, J. E., and Yurko, R. J., "Direct Determination of Wear Metals in Used Lubricating Oils by Arc Emission Spectroscopy Using a Novel Rotating Disc Electrode," presented at the 37th Pittsburgh Conference and Exposition on Analytical Chemistry and Applied Spectroscopy, Atlantic City, N.J., 12 March 1986.

[*8*] Yurko, R. J., Gilbert, T. R., and Sandoval, J. E., "Direct Analysis of Wear Particulates by the Ashing Rotrode Technique," in *Proceedings,* Condition Monitoring '87, University College of Swansea, Wales, Pineridge Press, Swansea, Wales, U.K., 1987.

[*9*] Fassel, V. A., Petersen, C. A., Abercombie, F. N., and Kniseley, R. N., "Simultaneous Determinations of Wear Metals in Lubricating Oils by Inductively Coupled Plasma Atomic Emission Spectroscopy," *Analytical Chemistry,* Vol. 48, 1976, pp. 516–519.

[*10*] Kjer, T. E., "Wear Rate and Concentration of Wear Particles in Lubricating Oil," *Wear,* Vol. 67, 1981, pp. 217–226.

[*11*] Taken in part from Anderson, D. P., "A Quantitative Method for the Determination of Large Particles in Lubricating Oil Samples Using ICP Spectrometer," presented at the 1989 Condition Monitoring and Preventive Maintenance Conference, Atlanta, Ga., 1–4 May 1989.

[*12*] Anderson, D. P., "Condition Monitoring by Wear Metal Analysis Using ICP Spectrometry," in *Proceedings,* Condition Monitoring '87, University College of Swansea, Wales, Pineridge Press, Swansea, Wales, U.K., 1987.

[*13*] Brown, J. R., Saba, C. S., Rhine, W. E., and Eisentraut, K. J., "Particle Size Independent Spectrometric Determination of Wear Metals in Aircraft Lubricating Oils," *Analytical Chemistry,* Vol. 52, 1980, pp. 2365–2370.

Mary Carol Williams[1]

Elemental Analysis of Waste Oils by Inductively Coupled Plasma Mass Spectrometry

REFERENCE: Williams, M. C., "**Elemental Analysis of Waste Oils by Inductively Coupled Plasma Mass Spectrometry,**" *Modern Instrumental Methods of Elemental Analysis of Petroleum Products and Lubricants, ASTM STP 1109,* R. A. Nadkarni, Ed., American Society for Testing and Materials, Philadelphia, 1991, pp. 96–104.

ABSTRACT: Environmental regulations specify the analysis of recycled oil for arsenic, cadmium, chromium, lead, and total halogens. Waste oils for disposal must be analyzed for arsenic, barium, cadmium, chromium, lead, mercury, selenium, and silver. Inductively coupled plasma mass spectrometry (ICP-MS) is a multielement technique with sufficient sensitivity to determine most metals and some nonmetals at the parts-per-billion level. Direct aspiration of the sample after dilution in a suitable solvent eliminates the tedious and potentially dangerous acid digestion process and the loss of volatile elements. The heat generated in the plasma and the addition of oxygen in the gas streams aid the breakdown of complexes of metals and the destruction of the organic matrix. The detection limits in a multielement analytical batch for elements of high mass [>80 atomic mass units (amu)] approach 5 to 50 $\mu g/m^3$ for undiluted samples. Elements of lower mass suffer from isobaric or polyatomic interferences, which degrade detection limits.

KEY WORDS: inductively coupled plasma mass spectrometry, elemental analysis, recycled oil, waste oil

Recycled oil may be reprocessed, used to produce a fuel, or burned for energy. The *Federal Register* of 29 November 1985 [*1*] published the final rule for handling, storing, transporting, and reuse of recycled oil. These regulations set the following maximum allowable limits for metals and halogens: arsenic (5 ppm), cadmium (2 ppm), chromium (10 ppm), lead (100 ppm), and total halogens (4000 ppm). Waste oils [*2*] are considered hazardous at the following levels: arsenic (5 ppm), barium (100 ppm), cadmium (1 ppm), chromium (5 ppm), lead (5 ppm), mercury (0.2 ppm), selenium (1 ppm), and silver (5 ppm). United States Environmental Protection Agency (EPA) regulations specify the levels in parts per million (ppm), which may be in milligrams per kilogram or micrograms per cubic metre. Direct aspiration of waste oils eliminates the tedious and time-consuming acid digestion process. The potentially explosive mixture of petroleum products and oxidizing acids and the loss of volatile elements are serious drawbacks of the acid digestion technique. Acids sufficient to destroy the organic matrix often cause interference problems in the instrumental techniques. This is particularly true of sulfuric and perchloric acids.

Flame atomic absorption (FAA), inductively coupled plasma atomic emission spectroscopy

[1] Staff member, Health, Safety, and Environmental Chemistry, Los Alamos National Laboratory, Los Alamos, NM 87545.

(ICP-AES), and direct-current plasma atomic emission spectroscopy (DCP-AES) are the most commonly used techniques for analysis of directly aspirated oil. Flame FAA is a time-consuming single-element technique and produces variable results when particulates are present. Furnace atomic absorption has better detection limits and less variation due to particulates, but the smoke during ashing and atomization, the volatility of the analyte, and the matrix interferences severely limit the usefulness of furnace analyses. DCP-AES and ICP-AES are multielement techniques and provide good companion instrumentation to inductively coupled plasma spectrometry (ICP-MS), especially for low-mass (<80 amu) elements. However, the necessity of diluting the sample to provide a more uniform matrix match with the standards challenges the detection limits of these techniques for several environmentally regulated elements, especially arsenic, mercury, and selenium. Neither emission technique has the ability to screen for halogens in oils with standard commercial instrumentation. Keane and Fry [*3*] described the use of ICP-AES with a photodiode array detector to determine levels of fluorine, chlorine, bromine, iodine, and sulfur. LaFreniere et al. [*4*] reported detection limits of 50 μg/m^3 for chlorine and bromine, using an optical sampling orifice inserted directly into the argon inductively coupled plasma (ICP). Helium plasmas have been used for determination of chlorine and bromine in aqueous solutions [*5*].

The ICP-MS has detection limits in the low micrograms-per-cubic-metre (nanograms-per-millilitre) levels, multielement capability, and oxygen-assisted plasma. Determination of chlorine, bromine, and iodine is possible with minor operational adjustments. In addition, a simple-to-run semiquantitative mode allows rapid screening to eliminate complete analysis of many samples. Previous papers [*6,7*] dealing with organic solution analysis by ICP-MS have been concerned primarily with wear metals. Hausler [*6*] presented a discussion of isobaric and polyatomic interferences and data showing the feasibility of semiquantitative analysis. Hutton [*7*] found that the spray chamber temperature and sample uptake were important considerations in reducing the solvent load on the plasma. He suggested 8 to 10°C for the spray chamber, later verbally revised to −10°C, and a sample uptake of 0.5 mL/min. By reducing the aspirated solvent, the amount of oxygen needed concurrently is reduced. It is important to balance the percentage of oxygen properly in the nebulizer flow. Too much oxygen seriously degrades the orifice of the sample cone; too little allows a rapid buildup of carbon on the sampler and skimmer cones.

Experimental Procedure

Sample Preparation

Waste oil samples were diluted 25-fold (2 g in 50 mL) and 100-fold (0.5 g in 50 mL) in kerosene. The 25-fold dilution was selected as the minimum to match the matrix with the standards and to reduce possible interferences. The 100-fold dilution often brought the high-element concentrations into the range of the calibration curve. It also served as a check for interference problems. For example, if the argon-chlorine polyatomic (mass 75) interferes with arsenic, a monoisotopic element, the analyst should be alerted by the differences between the calculated concentrations of the two dilutions. Some matrix interferences can be corrected by off-line software [*8*].

An aliquot of each dilution was spiked with 1 mg/m^3 (1 μg/mL) of standard for 25 and 100-mg/kg spikes. Indium was used as the internal standard (approximately 0.5 mg/m^3) to compensate for drift. Working standards were prepared fresh daily from the following standards obtained from Conostan™ (Conostan Division, Conoco Oil Co., Ponca City, Oklahoma): 500 mg/kg S-21 (a blend of 21 organometallic standards) and separate 100-mg/

kg standards for arsenic, selenium, and mercury. The calibrating standards included a blank and 10, 50, 100, 250, 500, and 1000 $\mu g/m^3$. All the standards and samples were run in triplicate.

Mass Spectrometer

A VG PlasmaQuad ICP-MS (VG Isotopes, Winsford, England) was started manually, and, by adjusting the lens settings, the sensitivity and resolution were peaked for an aqueous solution containing magnesium, indium, and lead. Once the ICP-MS was optimized for an aqueous solution, the sampling probe was removed and air was aspirated into the spray chamber long enough to remove all traces of water. The detector section of the ICP-MS was shut down, and the torch was moved back to the ignite position before the kerosene aspiration so that oxygen could be regulated without carbon buildup on the sampler cone. The peristaltic pump was fitted with Viton (a fluoroelastomer) tubing, and the peristaltic pump speed was reduced to permit a kerosene uptake of 0.5 mL/min. Ketones, alcohols, and halogenated hydrocarbons cause deterioration of the pump tubing, so it is important to use nonpolar solvents. Kerosene was used as the diluent because it is less toxic and less volatile than xylene. The plasma radio frequency (RF) power was increased to 1600 W and the cooling flow to 15 L/min before the aspiration of the kerosene. At the first appearance of the green carbon monoxide (CO) emission in the torch, oxygen was slowly added until the green at the base of the plasma disappeared. The torch was moved forward and the oxygen was readjusted so that the carbon buildup at the sampler cone was eliminated. After the rest of the instrument setup was completed, the ICP-MS sensitivity and resolution were again adjusted. Table 1 compares the instrument settings for aqueous and kerosene optimization. The first 11 settings are indicators of the voltages applied to the various components of the sample interface and the lens stack. Adjustment of these voltages focuses or filters the ion beam to optimize the response signal. The lens stack settings for the aqueous and organic solvents are not significantly different. The higher voltages for the collector and extractor are a function of the changes in the composition and energy of the plasma. Higher

TABLE 1—*Comparison of aqueous solution and kerosene instrumental parameters.*

	Aqueous Solution	Kerosene
Cone bias	0	0
Collector	5.31	5.56
Extraction	0.580	0.765
Lens 1	5.12	5.16
Lens 2	4.29	5.16
Lens 3	4.94	4.51
Lens 4	3.80	3.80
Differential aperture	0.400	0.530
Front plate	4.90	5.08
Pole bias	4.38	4.38
Prefilter	5.0	5.0
Nebulizer flow, L/min	0.700	0.700
Auxiliary flow, L/min	0.05	0.05
Cooling flow, L/min	14.0	15.0
Oxygen flow, L/min	0	0.090
RF power, W	1350	1600
Reflected power, W	<5.0	18
Counts/s, mg/m^3 In	7.1×10^6	2.2×10^6

RF power is required for organic solvents to prevent extinguishing of the plasma. The cooling flow was increased to cool the hotter plasma, and the higher reflected power is typical of organic aspirations.

Data Collection

Samples were introduced automatically into the instrument with a Gilson 221 sample changer. The following default settings of the VG software were used for pulse counting: dwell time, 320 μs per channel; mass range, 25 to 209 amu; number of scans, 120; and number of channels, 2048. The sample uptake and wash times were 70 and 120 s, respectively. A linear regression through a Lotus 1-2-3 spreadsheet converted the integrated data (in counts per second) to concentration.

Results

Table 2 shows the concentrations (in milligrams per kilogram) of environmentally significant metals in a used motor oil, giving the mean and within-run relative standard deviation for triplicate aspirations of each dilution. Recoveries of the 25 and 100-mg/kg spikes are also listed. The overall recovery for the 100-mg/kg spike was 107%, and that for the 25-mg/kg spike was 104%. Since samples are diluted 25-fold and 100-fold, a spike is added to each dilution as a check for interferences. If the spike results fall outside the range of 85 to 115% for both dilutions, the results can be corrected by the method of standard additions. The reason for the low spike recovery for the 100-mg/kg barium spike is not known. Matrix interferences in aqueous solutions are documented [*9,10*]; however, the interferences from organic matrix components are not as well understood at this point. A more exhaustive study needs to be done. The memory effects of silver and mercury are serious, especially with organic solutions. This accounts, in part, for the high spike recoveries for these metals. Mercury is a difficult metal to determine by ICP-MS. It has seven isotopes (196, 198, 199, 200, 201, 202, and 204), so the mass/charge ratio for any one isotope is reduced. In addition, the ionization potential of mercury is high relative to that of most metals, so it does not ionize as efficiently. Both of these factors reduce precision and sensitivity. The ICP-MS is used to screen for approximate levels of mercury; quantitative analyses must be done by cold vapor atomic absorption.

A qualitative scan of an organic solvent diluted to contain 40.5 mg/kg of bromine and

TABLE 2—*Quantitative trace metal results (milligrams per kilogram) for a used oil after dilution with kerosene.*

Element, mg/kg	Mean, %	Relative Standard Deviation (n = 3)	% Recovery, 100 mg/kg spike	% Recovery, 25 mg/kg spike
As	1.16	6.98	108	106
Ba	5.13	1.91	75.9	90.7
Cd	1.58	6.23	117	105
Cr	1.17	6.98	106	108
Pb	55.3	2.00	94.7	88.3
Hg	<1.3	...	134	116
Se	<1.3	...	105	109
Ag	2.17	3.69	113	110

34.5 mg/kg of iodine is shown in Fig. 1. The typical background species are labeled. Bromine and iodine peaks are easily distinguishable from the background. A semiquantitative scan includes a blank and a known standard in addition to the unknown sample. The integrated counts of six elements across the mass range produce a matrix-related plot of counts versus mass/charge. Calculations that include the response of each element to the internal standard and the matrix-related plot, and the database corrections for isotopic abundance and ionization, give semiquantitative results that are within a factor of five for complex organic matrices and a factor of two for well-defined organic solutions. When the regulatory limits are in the 1 to 4000 g/kg range, the ±500% results at the micrograms-per-kilogram level are adequate. The quantitative analysis of the sample by ICP-MS after dilution is suggested for individual elements that are close to the regulated limits; conventional methods are recommended when unresolved interference problems are present. Since most interferences are positive, because of the additive effects of matrix polyatomic species, a micrograms-per-kilogram result for arsenic, for instance, is well below the 5 ppm limit for either used or waste oils. Lead is the only element that commonly exceeds the regulatory limit, and it is easily reanalyzed by flame atomic absorption.

Quantitative results for halogen analyses suffer from both poor precision and inaccuracy. Organic solvents have a longer memory than aqueous solutions for all elements. This is due in part to the slower aspiration rate (0.5 mL/min for organic solvents versus 1.0 mL/min for aqueous), which does not allow as fast a rinse of the sample introduction tubing. There may

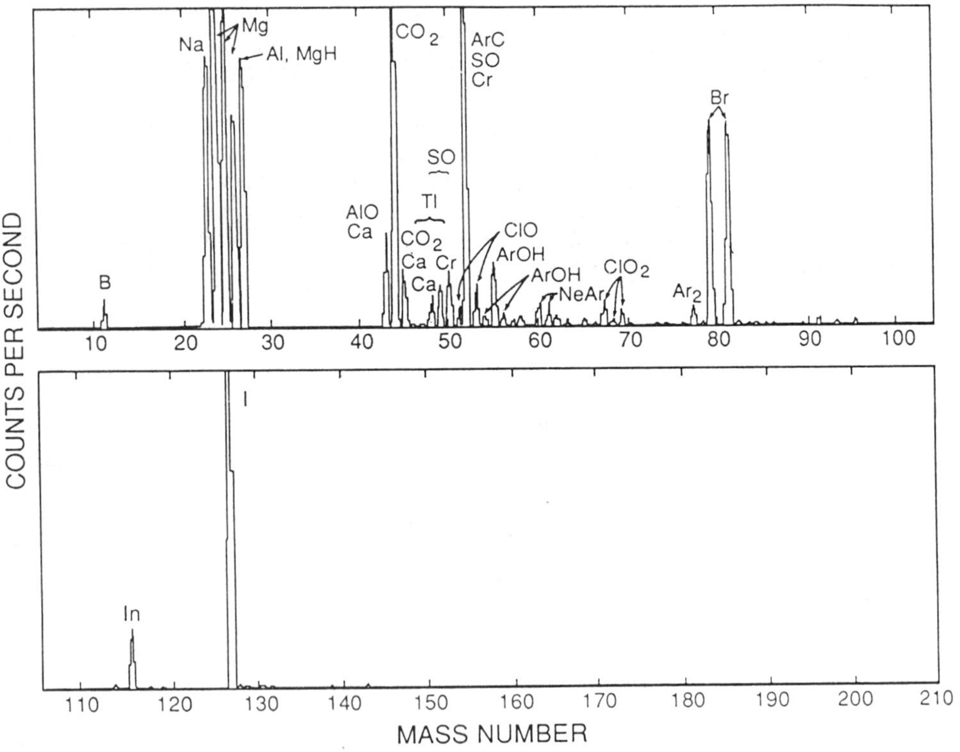

FIG. 1—*Qualitative scan of an organic solvent indicating background species and added bromine and iodine.*

also be a stronger interaction between the plastic tubing and the organic solvent than there is with an aqueous solvent. This effect was particularly noticeable for silver and mercury and for all three halogens. Quantitative analyses for chlorine, bromine, and iodine require a more detailed study.

Used oils can be rapidly screened to determine whether they can be recycled or must be disposed of as hazardous wastes. The heat capacity is part of the routine analysis requested for used oils. The solution resulting from the calorimetric bomb must be corrected for sulfur content. This is routinely done by ion chromatography (IC), which also measures the fluorine

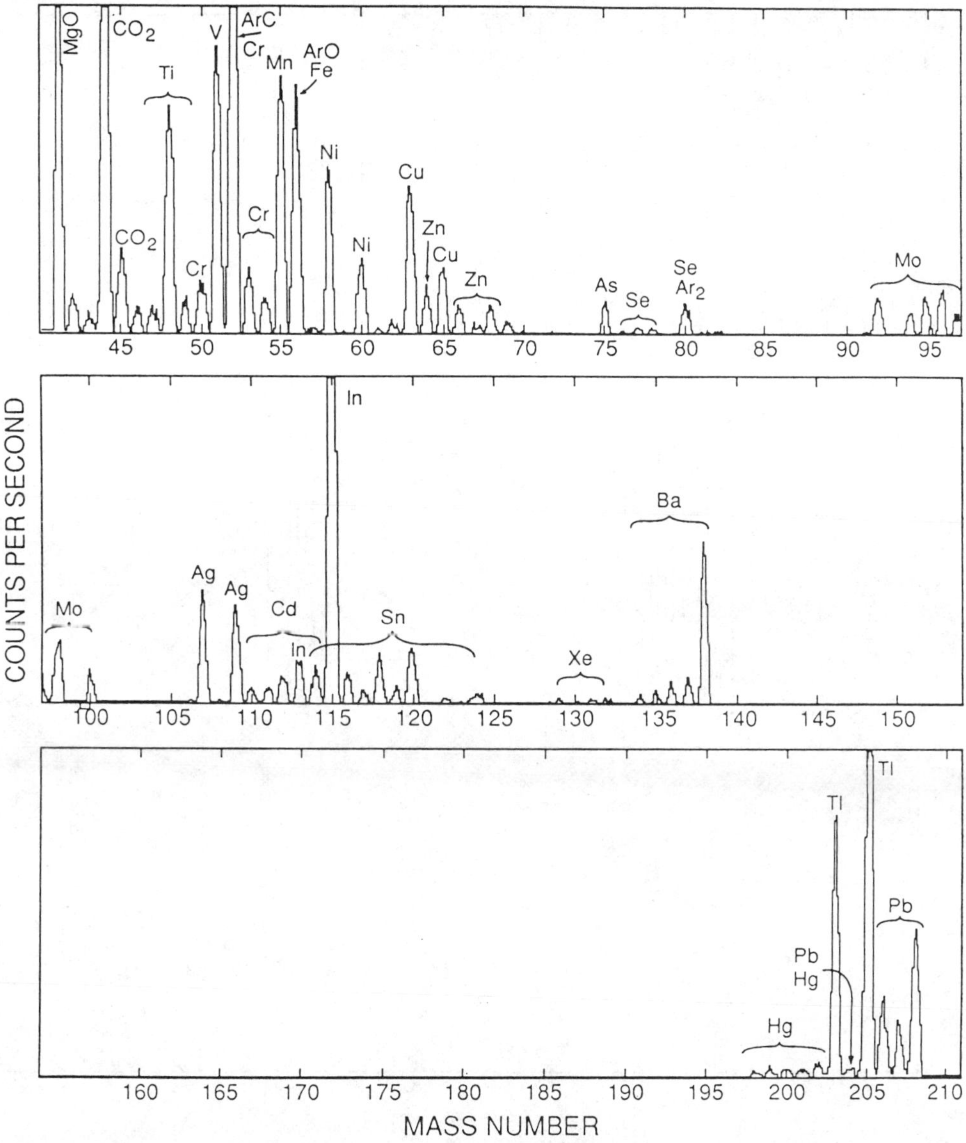

FIG. 2—*Scan of a mixed metal standard (0.5 ng/mL) indicating metals and background species.*

and chlorine content. However, a special column, eluent, and detector are needed to separate the bromine and iodine peaks. Since fluorine has such a high ionization potential, it cannot be determined as a positive ion by ICP-MS, and chlorine is in a mass range that is typically skipped by the VG software because of high background interferences. Fortunately, the combination of the ICP-MS and IC instruments allows quick screening without undue modification of either instrument. Trace metals, bromine, and iodine are determined by ICP-MS and chlorine and fluorine by IC in less than a day.

Figure 2 is a labeled plot of a 0.5 $\mu g/m^3$ (0.5 ng/mL) mixed standard that is typically run as a check standard to monitor semiquantitative analyses. A careful visual inspection of

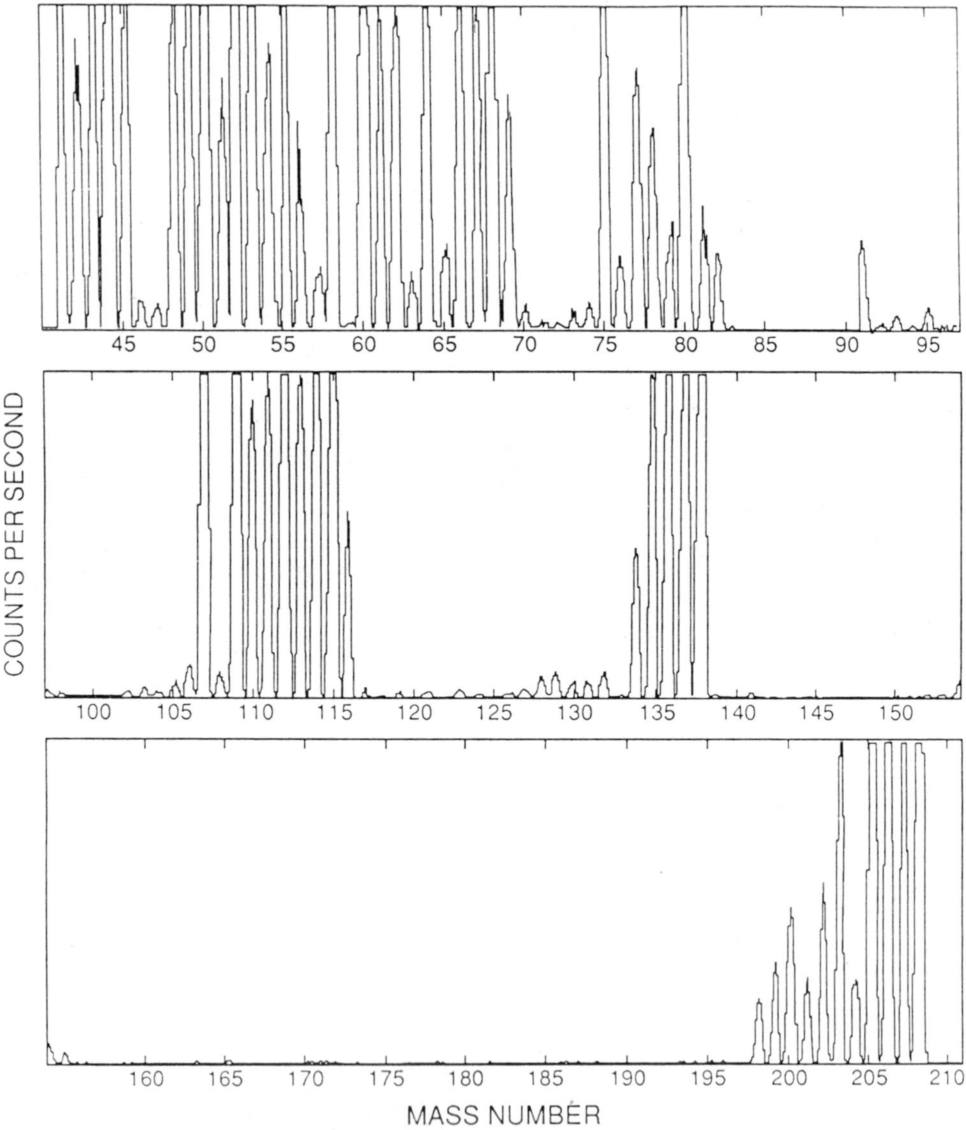

FIG. 3—*Hazardous waste scan.*

similar mass numbers in the plot of a used oil sample in Fig. 3 reveals it to be a hazardous waste. The mercury peaks in the unknown samples are much higher than the standard because the intensity of each mercury isotope is greater. Since the standard is 0.5 μg/m³ and the regulated limit is 0.2 μg/m³, the sample exceeds the limit for mercury. Arsenic (mass 75), cadmium (111, 114), chromium (53), lead (207, 208), and selenium (78, 82) appear to be much higher than the 0.5-g/m³ standard. The silver level is clearly low, since the two silver isotopic (107, 109) signals are small. Although the barium (134 to 138) peaks are much higher than the standard, the high regulatory limit (100 ppm) makes it more difficult to ascertain if the barium exceeds the limit. The iodine (127) and bromine (79, 81) levels are well below the 4000 ppm limits, but final determination will depend on the additive contributions from chlorine and fluorine. By converting the spectra from counts per second to log mode, the semiquantitative data allow the analyst to determine which elements must be quantitatively analyzed and what the most appropriate dilutions are. This semiquantitative information can provide an estimate about the final disposal of used oil in the time it takes to pump 5000 gallons from a storage tank into a tanker.

Conclusions

ICP-MS is a valuable tool for the determination of environmentally significant elements. The rapid screening of the semiquantitative mode allows the analyst to estimate the dilution levels and degree of accuracy required for quantitative analysis. Samples with low micrograms-per-cubic-metre (parts-per-billion) levels can be reported as less than a reasonable level; at the author's laboratory, this is usually two orders of magnitude below the limit of concern. Broader confidence limits (±50%) can also be associated with concentrations well below regulatory limits, but more accurate analysis is required for concentrations very close to these limits. In categorizing a sample as a hazardous waste, one must be sure to avoid either false positive or false negative results. ICP-MS screening provides a mechanism for making judgments for the complete quantitative analysis of each element, such as the type of matrix, potential interferences, range of the calibration curve, and dilution of the sample. As a quantitative technique, it lacks the precision and accuracy that are typical of results of aqueous solutions analyzed by ICP-MS or higher element concentrations in organic solutions determined by ICP-AES or DCP-AES. This is due, in part, to the matrix interferences. Combination species are possible from the air (hydrogen, nitrogen, oxygen), from the matrix (carbon, halogens, sulfur, phosphorous), and from the plasma gases (argon, krypton, xenon). As with any new technique, experimentation with increasingly complex samples will produce a better understanding of the analytical problems and provide guidance in developing the techniques to solve them. Presently, ICP-MS is at this stage of development. Although it has good potential for analysis of organic solutions, it also has analytical problems that need to be solved.

References

[*1*] U.S. Environmental Protection Agency, *Federal Register,* Vol. 50, No. 230, 1985, pp. 49249-49251.
[*2*] U.S. Environmental Protection Agency, Title 40, *Code of Federal Regulations,* Part 261, July 1987.
[*3*] Keane, J. M. and Fry, R. C., *Analytical Chemistry,* Vol. 58, No. 4, 1986. pp. 790-797.
[*4*] LaFreniere, B. R., Houk, R. S., Wiederin, D. R., and Fasel, V. A., *Analytical Chemistry,* Vol. 60, No. 1, 1988, pp. 23-26.
[*5*] Chan, S. K. and Montaser, A., *Spectrochimica Acta,* Vol. 40B, Nos. 10-12, 1985, pp. 1467-1472.
[*6*] Hausler, D., *Spectrochimica Acta,* Vol. 42B, Nos. 1/2, 1985, pp. 63-73.

[*7*] Hutton, R. C., *Journal of Analytical Atomic Spectrometry,* Vol. 1, 1986, pp. 259-263.
[*8*] "Inductively Coupled Plasma–Mass Spectrometry," Method 6020 CLP-M, Version 4.0, U.S. Government Printing Office, Washington, D.C., 1989.
[*9*] Tan, S. H. and Horlick, G., *Applied Spectroscopy,* Vol. 4, No. 4, 1986, pp. 445-460.
[*10*] Tan, S. H. and Horlick, G., *Journal of Analytical Atomic Spectrometry,* Vol. 2, 1987, pp. 745-763.

F. C. McElroy[1] *and J. M. Mulhall*[1]

Precious Metal Assay Analysis of Fresh Reforming Catalyst by X-Ray Fluorescence Spectrometry

REFERENCE: McElroy, F. C. and Mulhall, J. M., **"Precious Metal Assay Analysis of Fresh Reforming Catalyst by X-Ray Fluorescence Spectrometry,"** *Modern Instrumental Methods of Elemental Analysis of Petroleum Products and Lubricants, ASTM STP 1109,* R. A. Nadkarni, Ed., American Society for Testing and Materials, Philadelphia, 1991, pp. 105–117.

ABSTRACT: Precious metal analysis of fresh reforming catalysts are typically performed by both the catalyst manufacturer and buyer to arrive at a financial settlement on the quantity of metal in each lot of commercial catalyst. Traditional assay methods involve a variety of fire assay or wet chemical acid digestion schemes coupled with gravimetric, colorimetic, or titrimetric measurement for precious metals. Methods must have sufficient precision and accuracy to afford interlaboratory agreement of within one half of one percent relative between the catalyst supplier and purchaser. To meet this requirement many laboratories rely on classical methods. Unfortunately these procedures are labor intensive and time consuming.

X-ray fluorescence has the inherent instrument precision to achieve typical intralaboratory precision of 0.5% RSD on a wide variety of elements and numerous sample types. We have developed an X-ray fluorescence method for the assay quality analysis of fresh reforming catalysts containing platinum, rhenium, and iridium. This method was applied to numerous samples over the past five years.

The XRF method uses chemically assayed fresh catalysts as comparison standards to quantify the X-ray fluorescence response on unknown samples. Samples and standards of the same matrix and nominal elemental concentrations are ground, dried overnight, and then allowed to air equilibrate. Standards and samples are pressed into briquettes with alumina cups in a pellet press to 20 tons. Net counts for each element are collected at the appropriate wavelengths for both standards and samples. Analysis is performed on a minimum of three sample briquettes and four briquettes for each of three standards. The ratio of the net counts to the precious metal concentration on each standard is averaged to obtain a calibration factor. Platinum alone and platinum in combination with other promoter precious metal concentrations are calculated by taking the average sample net counts for each element and dividing by the appropriate calibration factor. Intralaboratory and interlaboratory results are presented showing the applicability of X-ray fluorescence as an assay quality procedure for precious metal analysis.

KEY WORDS: assay analysis, X-ray fluorescence, platinum, iridium, rhenium, precious metals, reforming catalyst

Alumina-based catalysts containing platinum and promoter metals such as rhenium and iridium are used in the petroleum industry for catalytic reforming, an important process in the production of high octane motor gasoline [*1*]. Concentrations of these elements are typically in the range of 0.3 to 0.7 wt%. Because tens of thousands of pounds of catalyst are used in commercial units, the monetary value of the precious metals requires accurate

[1] Exxon Research and Engineering Company, Corporate Research–Analytical Sciences Laboratory, Linden, NJ 07036.

analyses. This is done to ensure metal accountability when catalysts are purchased from the manufacturer and when reclaimed at a reclaimer. Typically each production lot is analyzed by both the manufacturer and the buyer or the buyer's representative. At Exxon we perform all of our own metal assay analyses.

Most laboratories employ classical gravimetric or colorimetric methods in order to meet the requirement of interlaboratory results being within 1% of each others troy ounce value [*2*]. Colorimetric methods rely on standards prepared from gravimetric determination on primary materials. These are then used as calibration standards in the colorimetric analysis [*3*]. Because both gravimetric and colorimetric methods are time consuming and labor intensive a new method using X-ray fluorescence spectroscopy was developed for the direct analysis of platinum, rhenium, and iridium on alumina-based catalysts.

Sampling

The protocol shown in Fig. 1 is followed to determine the quantity of precious metal in each manufactured lot of catalyst. First a sample representing nominally 10 000 lb of catalyst is obtained at the manufacturing site. Assay analyses are performed by both the manufacturer and buyer. On an agreed upon date each mails results to the other party. If there is less than a 1% troy ounce difference, metal settlement is based on the arithmetic mean. If the results are outside the 1% splitting limit, either party may elect to reassay; if the results are still outside limits, assays are performed at a third-party laboratory to settle the analytical dispute. Representative sampling and accurate analysis are key requirements in effectively using this protocol to measure the precious metal content.

Even the best analytical method is only as good as the sample being analyzed. A well defined sampling protocol is very important, since a 1 lb sample will represent a nominal lot size of 10 000 lb. Sampling is done using automatic equipment designed to sample, at periodic time intervals, the entire contents of a falling catalyst stream (Fig. 2). The bulk sample is then blended and representative samples removed for moisture content determination. The catalyst is produced in extrudate form. To ensure a representative sample

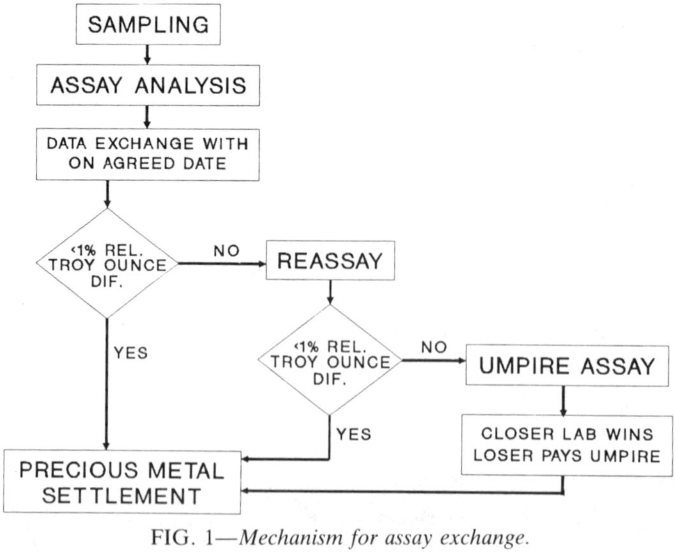

FIG. 1—*Mechanism for assay exchange.*

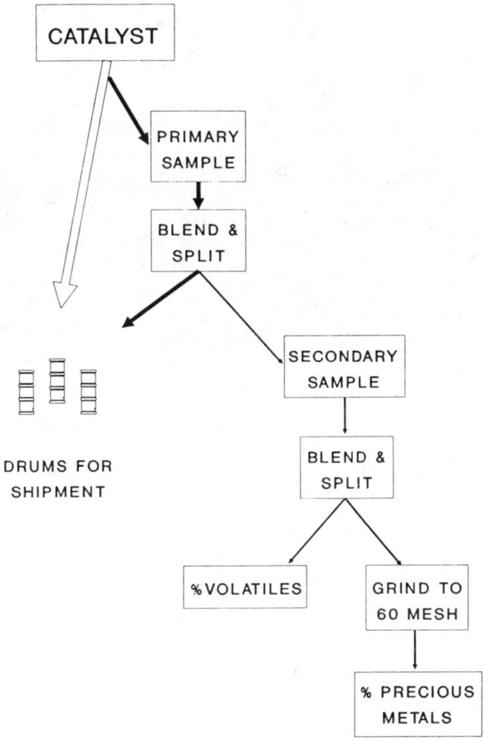

FIG. 2—*Catalyst sampling protocol.*

for metal assay, the extrudate is ground to pass 60 mesh. The ground material is then blended and representative samples removed. The original extrudate is analyzed for percent volatiles to convert the shipment lot weight to an anhydrous basis. The ground material is analyzed for the percent metals converted to an anhydrous basis. Analyses of the extrudate and ground samples are required to calculate the anhydrous troy ounce metal content for the shipment lot. Both the manufacturer and the buyer perform these analyses on separate samples. Additional samples are sealed and retained in case referee laboratory analyses are required.

Analysis

Requirements of an assay procedure are summarized in Fig. 3. Both the original extrudate and ground samples from each production lot are analyzed for the percent volatile materials. Metal assay analysis is performed on the ground sample. Analytical procedures must provide intralaboratory precision of about 0.5% RSD.

Loss-on-Ignition Analysis

Because grinding can alter the water content of a sample, both the extrudate and ground samples must be analyzed for volatiles to correct the measured precious metal content of the ground sample to the same anhydrous basis as the extrudate. Since the interlaboratory comparisons are made on a troy ounce basis, the temperature for determining the volatile

- PROCEDURES TO OBTAIN REPRESENTATIVE SAMPLES
- MEANS TO CORRECTS WEIGHTS TO DRY BASIS
- METHODS THAT ARE PRECISE AND ACCURATE
  - 0.5% RSD INTRALAB
  - 1.0% TROY OUNCE MEAN INTERLAB
- ANALYTICAL OPTIONS
  - GRAVIMETRIC
  - COLORIMETRIC
  - X-RAY FLUORESCENCE
  - INDUCTIVELY COUPLED PLASMA

FIG. 3—*Precious metal assay analysis.*

content is not critical as long as all the water is removed and a constant weight is obtained. Since alumina in the alpha form does not absorb water, we prefer to calcine at temperatures that will easily convert the gamma alumina to the alpha form. This is done in less than 1 h by calcining at 1427°C (2600°F). At these temperatures virtually all the rhenium is lost. Therefore it is important that one takes a separate ground sample for metal analysis.

Other temperatures for determination of volatile losses yield the same troy ounce metal content for the production lot. A lower calcination temperature results in less loss of volatile material with a lower anhydrous precious metal percentage, but a higher anhydrous weight for the production lot. The troy ounce content for the lot is calculated by multiplying the anhydrous precious metal percentage by the anhydrous weight of the shipment and by the conversion factor for pounds to troy ounces (14.5833). In this way the troy ounce content of each manufactured catalyst lot is independent of the calcination temperature. However, the same temperature must be used on both the extrudate and ground samples.

Assay Analysis for Metals (Traditional Methods)

Traditional assay methods involve a variety of fire assay or wet chemical acid digestion schemes coupled with gravimetric or colorimetric measurement [*4*]. Gravimetric procedures require a series of precipitation strategies to isolate and weigh the pure metal. Colorimetric methods usually require that the metal is isolated in fractions and then typically reacted with stannous chloride and the visible absorbance compared with a set of standards. Both of these approaches are time consuming and manpower intensive.

Experimental Procedure

Instrument Measurement Conditions

Wavelength dispersive X-ray fluorescence (WDXRF) instrumentation has a inherent counting precision of 0.2 to 0.3% due to the stability of the X-ray tube and the detector systems [5]. All analyses were done on a Philips Electronic Instruments PW-1400 wavelength dispersive instrument equipped with a chromium X-ray tube. The optical path was helium and the analyzing crystal was LiF (200). Both a flow proportional counter (90% argon + 10% methane) and a scintillation counter were summed for the detection of the fluorescence radiation.

The instrument operating conditions listed in Table 1 were selected to provide an intra-laboratory analyte measurement precision of at least 0.5%. To do this, counting times on

TABLE 1—*X-ray spectrometer operating conditions.*

Element	Line	Measurement Angle (°2θ)			Time (s)	
		Peak	Bkgd.	Bkgd.	Peak	Bkgd.
Pt and Pt/Re:						
Pt	$L\alpha_1$	38.050	37.050	42.680	80	24
Re	$L\beta_1$	35.820	34.320	37.050	80	24
Pt/Ir:						
Pt	$L\alpha_1$	38.050	37.050	40.210	80	24
Ir	$L\alpha_1$	39.210	37.050	40.210	80	24
X-ray tube	Chromium target					
Voltage	60 kV					
Current	50 mA					
Crystal	LiF (200)					
Collimator	Fine					
Path	He					
Detector	Flow and scintillation					

the peak and background were selected to provide adequate counting statistics to ensure good measure repeatability on each sample specimen.

The selection of the X-ray fluorescence analyte line to measure was based on a combination of the intensity of the line and its relative freedom from possible interferences. Because one should excite the element at 3 to 5 times the excitation potential of the analyte, only the *L* lines were considered for the measurement of platinum, rhenium, and iridium. Normally the $L\alpha$ lines are the most sensitive; however, there is a line overlap of the rhenium $L\alpha_1$ (41.69°2θ) with the Zn K_α(41.80°2θ). To eliminate any possibility of an interference, the rhenium $L\beta_1$ line was selected. A typical wavelength scan over the platinum, iridium, and rhenium analyte region is shown in Fig. 4.

Standard Preparation

Separate sets of triplicate standards were developed for each catalyst type analyzed. The analyte concentration of the sample is within 10% relative to that of the standards. Platinum, rhenium, and/or iridium concentrations were measured by classical gravimetric or stannous chloride colorimetric methods. In all cases agreement between the values on the standards were within the 1% relative value obtained by the manufacturer. Since the alumina based catalyst contains nominally the same elemental base matrix, any elements affecting the measured XRF intensities on the standards would also affect the samples. Therefore it is important to analyze a set of standards for each catalyst type manufactured using gravimetric or colorimetric methods.

Sample Analysis

The procedure used in the analysis of the standards and samples is summarized in Fig. 5. All samples are ground to pass 60 mesh if not already received in this condition. Approximately 20 g of each sample and 20 g of each of three standards are placed in 100 mL silica dishes and dried at 200°C overnight in a drying oven. Samples and standards are then removed and allowed to air equilibrate at room temperature for at least 48 h. This is done

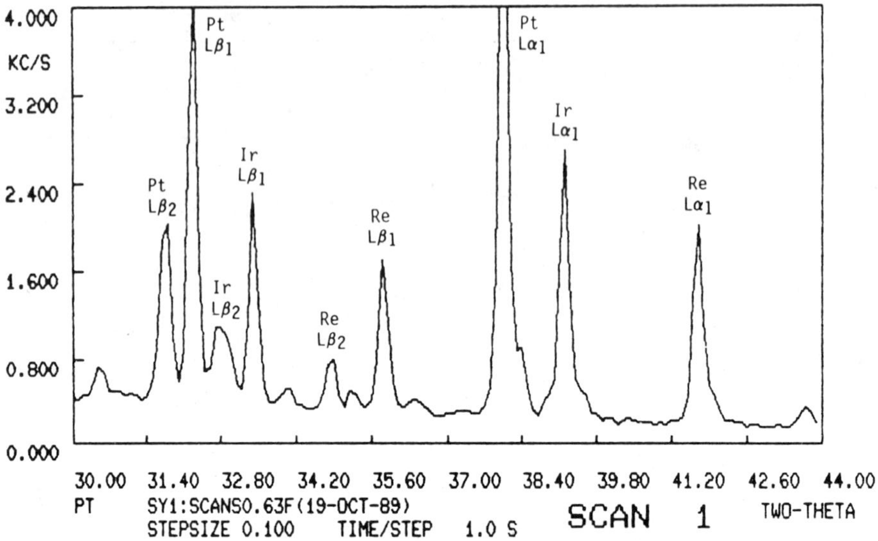

FIG. 4—*Wavelength scan of catalyst briquette.*

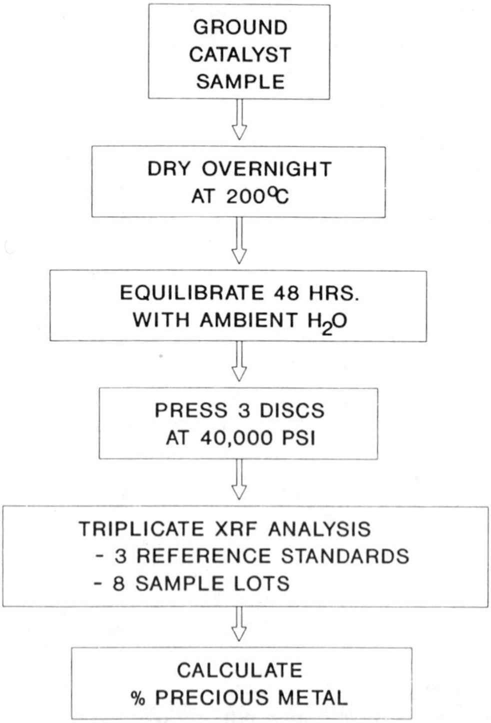

FIG. 5—*XRF catalyst assay method.*

so that the samples and standards absorb the same amount of water. Triplicate briquettes for each sample are prepared by filling 30 mm thin-walled alumina cups with excess air-equilibrated ground catalyst (approximately 3 g) and leveled with a spatula. The cap is inserted into a press and a force of 8000 lb applied. Gradually the pressure is increased to 40 000 lb over 15 to 20 s. This pressure is maintained for 30 s. The pressure is slowly reduced to 8000 lb over 15 to 20 s and then the pressure is released and the briquette removed. A minimum of four briquettes for each of three standards is prepared for each of the different catalyst matrices being analyzed.

Four briquettes for each of the three standards are used when analyzing one to eight samples. Standards and samples are analyzed using the instrument conditions specified in Table 1. Analysis of each group of samples are bracketed by the set of three standards and the sequence repeated for all three briquettes on each sample. The order for analyzing up to eight samples is: standards, samples, standards, samples, standards, samples, standards.

The concentration of the measured analyte is obtained by first calculating calibration factors for each of the three standards. This is done by dividing the known anhydrous concentration by the net peak intensity (peak − background) of the analyte element. An average calibration factor for each analyte (Pt, Re, or Ir) is obtained for each set of standard replicates. The calibration factor for each of the three standards is averaged to obtain one factor for each analyte. The concentration of the analyte element in the unknown sample is calculated by multiplying the net peak intensity by the analyte calibration factor. The average concentrations for the replicate samples analyzed is used to obtain the concentration of the analyte.

Moisture Correction

Since the catalyst is gamma alumina, one must control the water content of the standards and the samples. To eliminate any water effect on the X-ray fluorescence intensity, all samples and standards are dried at 200°C overnight and then allowed to air equilibrate in the laboratory for at least 48 h before analysis. Since the standards are analyzed for the precious metals content based on the calcination temperature of 1427°C, there is no need to perform volatility losses on the samples.

Results and Discussion

Method Development

X-ray fluorescence (XRF) has the advantages of being precise, rapid, and requiring minimal sample preparation. It has a limitation in that the measured analyte intensities are dependent on the concentration of elements in the sample matrix. Matrix elements can absorb or enhance fluorescence radiation of the analyte element. If samples are not matrix matched to the standards, analyte results will be biased.

To relate XRF intensities with analyte concentrations one can measure the elements causing the matrix effect and apply a mathematical correction, add an internal standard that would experience the same effect as the analyte, or utilize standards of the same matrix and metal loadings as the samples. Compensating for the matrix effects of different elemental compositions would require measuring the interfering element. This would also include the water content of the standards and samples. The use of an internal standard element such as gold could compensate for matrix effects, but this would require additional sample handling to add the gold. Since the matrix of a given fresh catalyst type remains almost constant with respect to the precious metal loading and other base elements, standards can easily be developed for use in instrument calibration. In this way analysis is performed by ratioing measured XRF intensities of the matrix matched standards to the unknown samples. Sample preparation is minimal.

A number of factors beside matrix effects must be considered in developing an accurate XRF method. Instrument operating conditions including analyte line selection must be defined. This was discussed in the Experimental Procedures section. Because of the requirement for high precision and accuracy, specimen preparation must also be considered. These include X-ray penetration depth, particle size distribution, analyte line selection, excitation, and measurement conditions.

Sample Thickness

Because of the relatively light aluminum matrix one should ensure that the X-ray penetration depth is sufficient so that not just the surface is being analyzed. On the other hand, the sample should be infinitely thick to eliminate sample thickness variations. A series of experiments were conducted by placing ground platinum/rhenium catalyst in XRF cups fitted with Mylar covers. XRF intensity was plotted as a function of sample weight. Based on the results illustrated in Fig. 6, a minimal amount of catalyst (>1.0 g) provides a constant fluorescence signal indicative of an infinitely thick sample.

Particle Size Effects

Sample thickness experiments showed that minimal sample is needed for obtaining a constant fluorescence intensity. When dealing with powders one must be sure that the

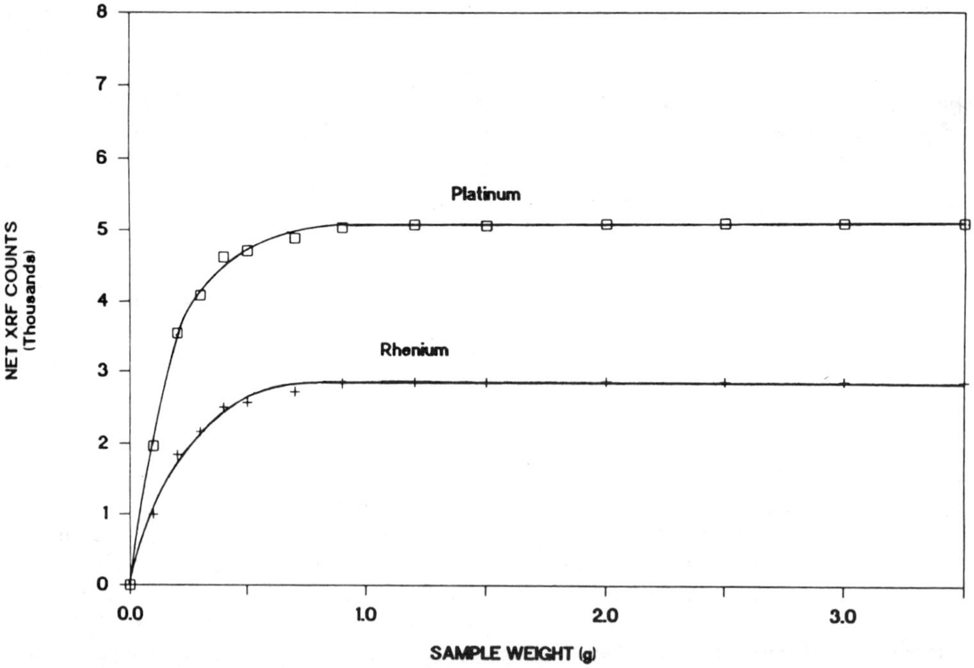

FIG. 6—*XRF intensity versus sample weight.*

penetration depth is sufficient to obtain a constant fluorescence signal independent of particle size [6]. This is accomplished by analyzing samples ground at different grinding times to different mesh sizes.

Figure 7 illustrates the comparison of XRF platinum intensity as a function of grinding time. Data on both powders and briquettes are presented. After grinding for 30 s using a puck and ring grinder and briquetting the specimens, there was no change in platinum X-ray fluorescence intensity. The powder specimens required much longer grinding times to yield a finer mesh size before the fluorescence signal stabilized. In practice briquetting specimens is a relatively simple procedure, since no binders are needed and there is less segregation possible due to the powders settling [6]. Based on this experiment and since all samples are received ground to pass 60 mesh, the specimen preparation procedure includes briquetting both standards and samples.

Intralaboratory and Interlaboratory Precision

Intralaboratory repeatability on triplicate samples analyzed over the past five years were compiled and the % RSD is graphed in Figs. 8, 9, and 10 for platinum, rhenium, and iridium respectively. The data are summarized in Table 2. In almost all cases the target of 0.5% RSD on intralaboratory analyses was achieved for platinum and iridium. In the case of rhenium intralaboratory precision was almost always better than 1.0% RSD.

To measure the interlaboratory reproducibility, results of exchanged data were compiled between us and the manufacturers over five years. These are plotted in Figs. 11, 12, and 13 for platinum, rhenium, and iridium respectively. Most platinum exchange data are close to the defined 1% splitting limit with good distribution around the mean. Although the

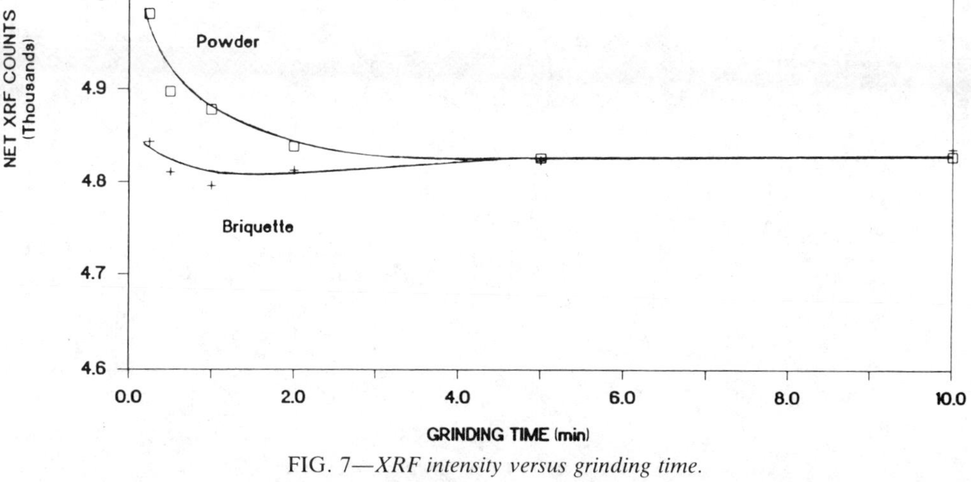

FIG. 7—*XRF intensity versus grinding time.*

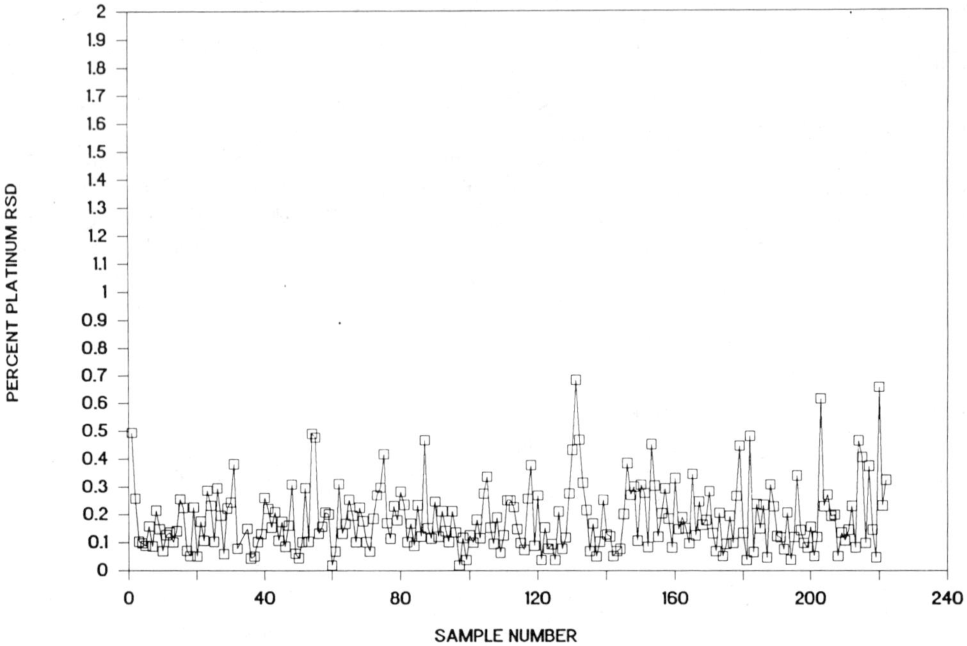

FIG. 8—*Intralaboratory platinum precision.*

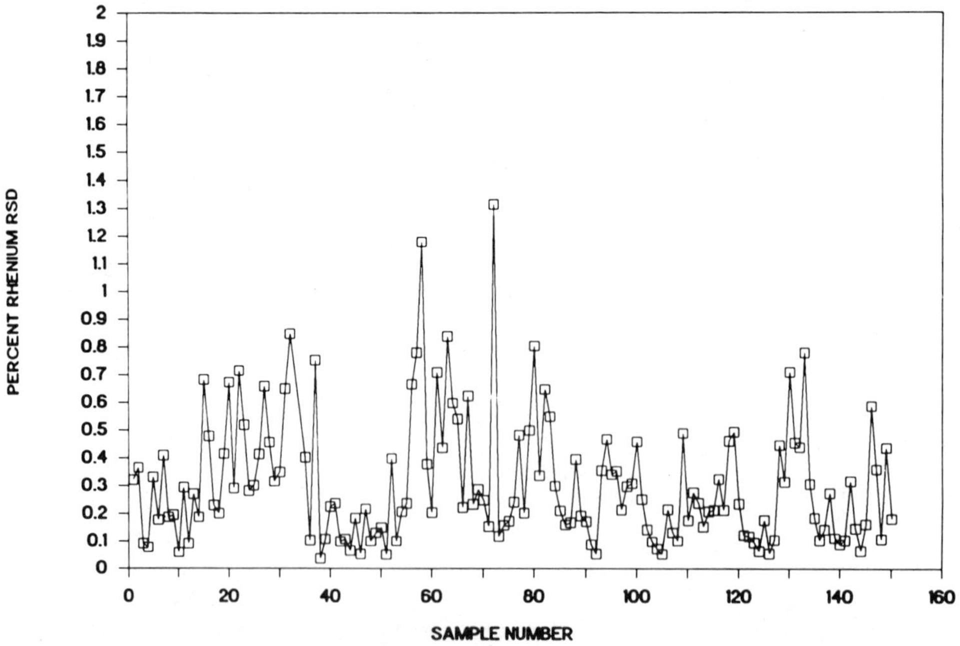

FIG. 9—*Intralaboratory rhenium precision.*

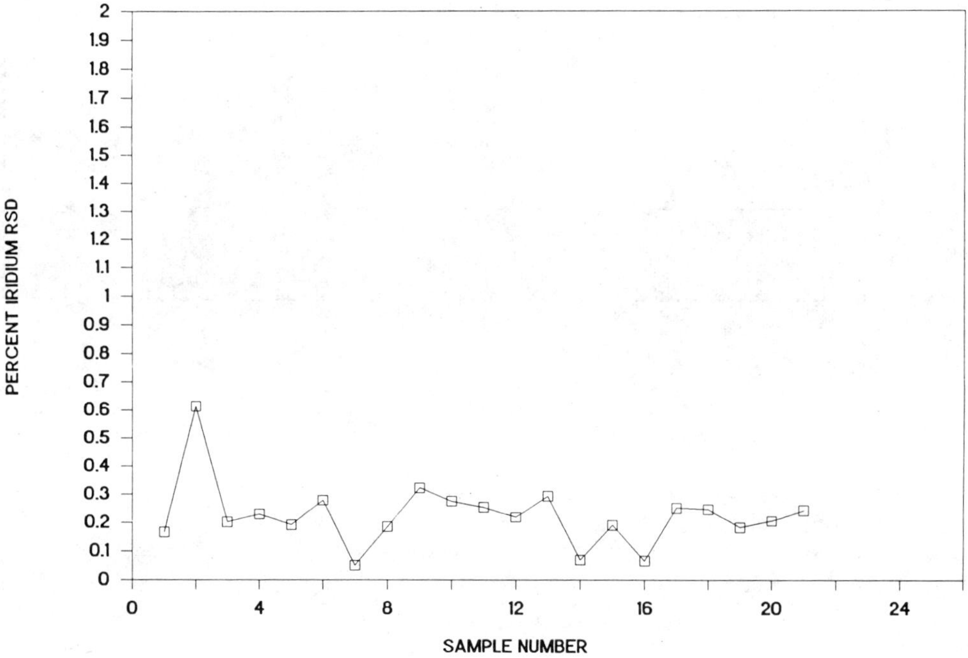

FIG. 10—*Intralaboratory iridium precision.*

iridium appears to show a small positive bias, the average result is within the 1% splitting limit. Rhenium shows a much wider spread in the interlaboratory reproducibility, but the overall mean is still within the 1% splitting limit. These results are also summarized in Table 2.

Although relative differences in platinum and rhenium concentrations will cause interelement matrix effects, the magnitude of these effects are small. No correlation was found in comparing the absolute platinum concentrations to the interlaboratory differences in rhenium. This is consistent with theoretical calculations showing that a 10% relative change in absolute platinum concentration would result in less than a 0.001% change in rhenium concentration.

TABLE 2—*Summary of intralaboratory precision and interlaboratory reproducibility.*

	Intralaboratory[a]		Interlaboratory[b]		
Element	*n*	% RSD	*n*	Mean % Bias	σ
Pt	220	0.27	215	−0.03	0.81
Re	148	0.47	144	−0.70	1.18
Ir	21	0.31	21	0.33	0.68

[a] Repeatability based on triplicate analyses of each *n* sample.

[b] Mean % calculated for each *n* by dividing the difference in interlaboratory troy ounce value by the average value.

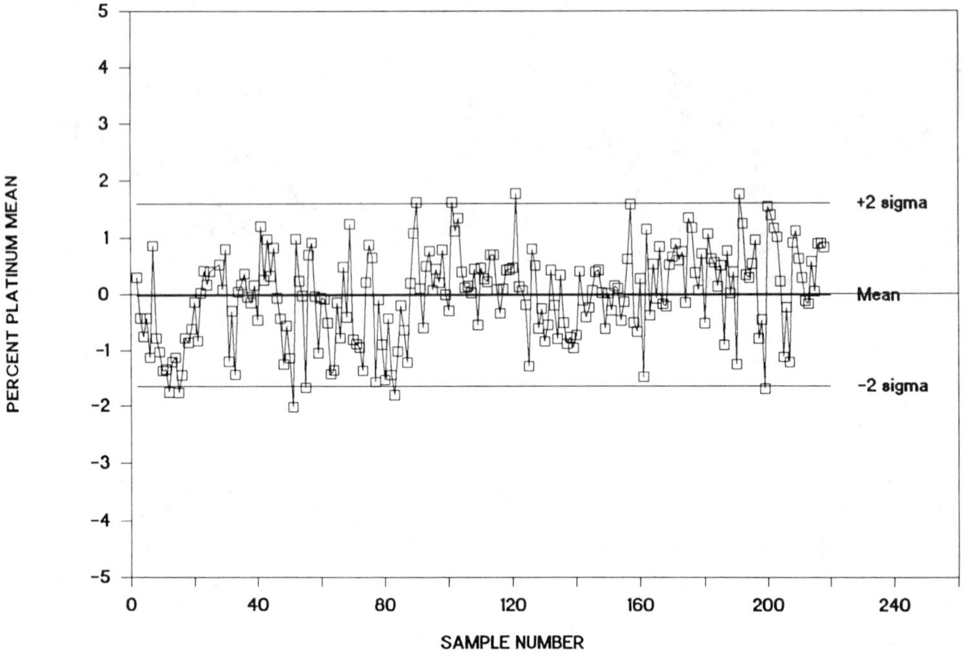

FIG. 11—*Interlaboratory percent platinum mean.*

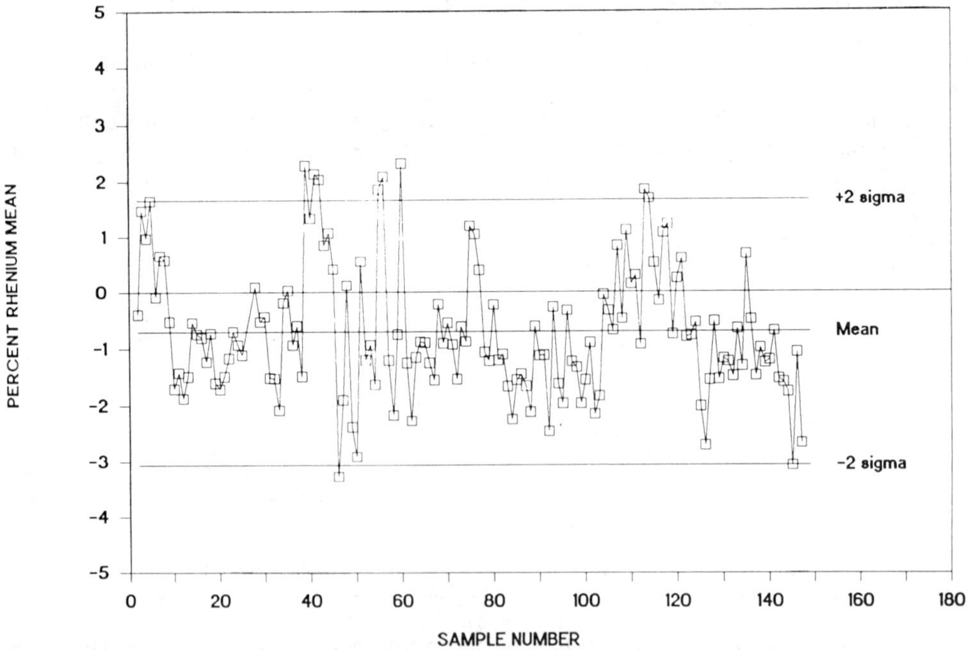

FIG. 12—*Interlaboratory percent rhenium mean.*

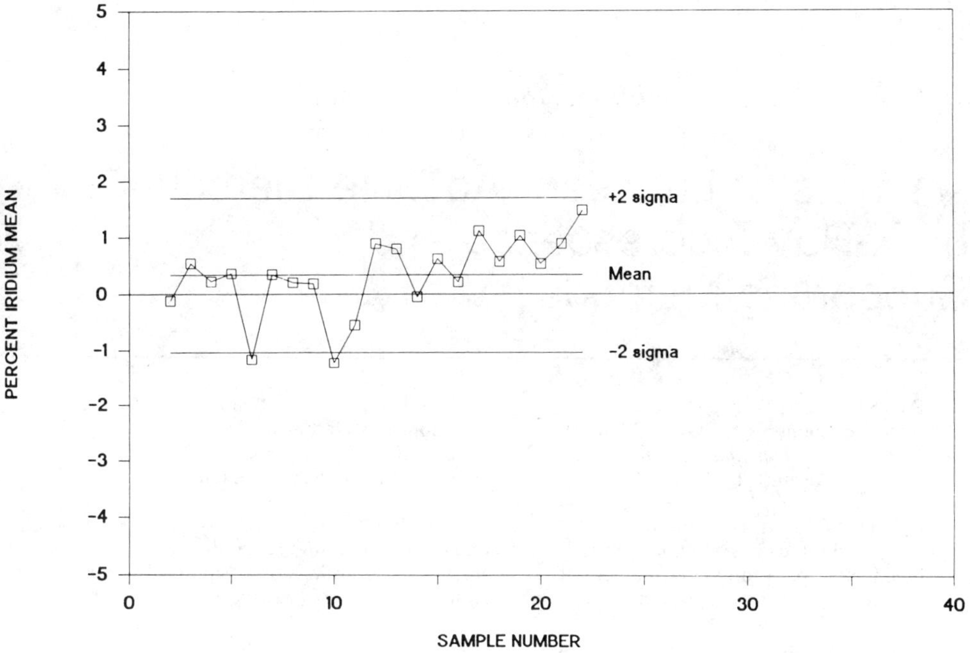

FIG. 13—*Interlaboratory percent iridium mean.*

Conclusions

Based on the excellent intralaboratory repeatability and the good interlaboratory reproducibility, we found that the X-ray fluorescence method is a cost-effective way to rapidly assay platinum, rhenium, and iridium in fresh reforming catalysts. Once standards are developed for the analyte element in the sample matrix, analyses of up to 17 sample lots can be completed in an overnight analysis.

References

[1] Sinfelt, J. H., *Bimetallic Catalysts, Discoveries, Concepts, and Applications,* Wiley, New York, 1983.
[2] Davies, A., "Methods and Problems in Platinum Analysis," in *Proceedings,* Seminar on Precious Metals Sampling and Analysis, S. Kallman, Ed., Met-Chem Research Inc., Boulder, Colo., 1989, p. 91.
[3] Beamish, F. E., *The Analytical Chemistry of the Noble Metals,* Pergamon Press, New York, 1966.
[4] Kallman, S. and Blumberg, P., "Applications of the Copper Sulfide Collection System," in *Precious Metals,* U. V. Rao, Ed., Met-Chem Research Inc., Boulder, Colo., 1986, p. 83.
[5] Jenkins, R., Gould, R. W., and Gedcke, D., *Quantitative X-Ray Spectrometry,* Marcel Dekker, New York, 1981.
[6] Bertin, E., *Principles and Practices of X-ray Spectrometry Analysis,* Plenum Press, New York, 1975.

J. R. Sieber,[1] S. G. Salmon,[1] and M. C. Williams[1]

Analysis of Lubricant Additive Elements by X-Ray Fluorescence and Supporting Methods

REFERENCE: Sieber, J. R., Salmon, S. G., and Williams, M. C., "**Analysis of Lubricant Additive Elements by X-Ray Fluorescence and Supporting Methods,**" *Modern Instrumental Methods of Elemental Analysis of Petroleum Products and Lubricants, ASTM STP 1109,* R. A. Nadkarni, Ed., American Society for Testing and Materials, Philadelphia, pp. 118–127.

ABSTRACT: At Texaco's Research Center, X-ray Fluorescence (XRF) is our workhorse method for the determination of additive elements in lubricating materials. Other analysis techniques are used to assist in method development by confirming compositions of standards and quality control samples, or by verifying analysis results from newly developed XRF methods.

Often it is impractical to make a determination of all analytes in a sample by XRF due to inherent limitations of X-ray fluorescence measurements and the complexity of calibrating for many analytes in a single method. Analyses of high volumes of samples can be made more efficient by omitting difficult or infrequently encountered elements. When a unique sample comes along, alternative analytical techniques are used to determine elements for which the X-ray spectrometer is not sensitive or not calibrated. These concentrations can be combined with theoretical influence coefficients to correct existing XRF calibrations for the effects of unanalyzed elements. A working example, sodium in research samples, is given. Another example, barium in oils, shows the serious shortcomings of this approach. Alternative techniques such as inductively coupled plasma atomic emission spectroscopy, atomic absorption spectrometry, sulfur determination by high temperature combustion with infrared detection, and wet chemical methods can be used to complete an analysis or to confirm the accuracy of XRF methods for determinations of additive elements in lubricants. Examples of our elemental analysis methods, how they relate to ASTM procedures, and how we use our techniques are described.

KEY WORDS: lubricants, additives, elemental analysis, standard test methods, X-ray fluorescence, gravimetric, inductively coupled plasma, calibration

Modern wavelength-dispersive X-ray fluorescence spectrometers are extremely sensitive and precise. Sample introduction is simply and rapidly performed. Calibration standards, especially for liquids, are readily available or easily prepared for custom applications. For these reasons, XRF is an excellent tool for the determination of additive element compositions in lubricants. At the Texaco Research Center, Beacon, New York, we employ X-ray fluorescence along with a collection of other elemental analysis techniques to analyze the wide variety and large numbers of research samples and production quality control samples generated by our laboratory and many of the Texaco blending and distribution terminals.

[1] Project Chemist, Senior Research Chemist, and Senior Chemist, respectively, Texaco Research and Development, Beacon, NY 12508.

Most of these samples fall into two categories of high volume work. The first and larger category is lubricant oils and additive packages. These may be novel performance additives, oils from performance tests, purchased ingredients, or finished products ready for the consumer. The second category is fuels from sales terminals, refineries, or our catalyst research groups. Methods have been developed to allow routine analyses of these types of materials. This paper will focus on the first category, lubricant oils and additive packages.

X-Ray Fluorescence Methodology

Definitions and Parameters

The analysis needs of our research personnel encompass the variety of samples mentioned above. Our approach to fulfilling these needs is to analyze the most frequently encountered elements in oils and fuels by X-ray fluorescence. The elements in Table 1 are present in performance additives and base stocks used to prepare research samples. Less frequently encountered elements include Na, Si, K, and Mo. Matrix elements for which our spectrometer is not sensitive are H, B, C, N, and O.

We are responsible for analyses of as many as 1000 samples per month in this category. To accomplish this task, we have equipped our Philips PW1404 with a 72 position sample changer. The spectrometer also has a Rh tube; LiF (200), Ge (111), and PE (002) crystals; and an ultrathin window proportional detector. Multiple regression calibration and calculation software plus quality control and data transmission features round out the system.

Calibration

The calibration approach chosen for this method was to employ a large number of standards to empirically modify theoretical interelement influence coefficients [*1*] to calibrate as wide a concentration range for each element as practicable. Theoretical alphas are calculated from first principles modeling of the physics of X-ray interactions with matter. Calculated coefficients were provided by the instrument vendor. The primary reason for choosing the mathematical approach to calibration was that we anticipated having to prepare and analyze 72 samples or more per day plus several dozen samples for other analyses. Additions of internal standards to every sample was deemed highly undesirable.

Standards were drawn from several sources to create a set of four dozen liquids. Thoroughly analyzed Texaco additive packages and base oils were used in neat form or diluted in mineral oil singly and in combinations. Commercially available single element standards

TABLE 1—*Comparison between concentration ranges present in lubricating oils and additives and ranges calibrated in X-ray fluorescence method.*

	Composition Range, wt%		Calibration Range, wt%	
Element	Min.	Max.	Min.	Max.
Barium	0	5	0.001	0.2
Calcium	0	20	0.005	1.0
Chlorine	0	3	0.003	0.4
Copper	0	0.5	0.0005	0.05
Magnesium	0	0.2	0.004	0.1
Phosphorus	0	10	0.005	1.0
Sulfur	0	30	0.01	2.5
Zinc	0	15	0.001	1.0

and a chlorinated paraffin were added to the additives and mixtures to prepare standards for the elements Ba, Mg, and Cl and to increase the variety of the mixtures. Standard Reference Materials (SRMs) from the National Institute of Standards and Technology were included in the set of calibrating standards. The specific SRMs chosen were 1618, 1622, 1634b, 1818 (I–V), and 1819 (I–V). Finally, di-*n*-butylsulfide was diluted in mineral oil to prepare standards as specified in ASTM Test for Sulfur in Petroleum Products (X-ray Spectrographic Method) (D 2622).

It was anticipated that we could not calibrate to include high concentrations of Ca, P, S, and Zn due to self-absorption of the analyte's X-rays. Therefore we surveyed the types and frequencies of samples submitted for additive element determination to find the most populated concentration ranges. We began with wider calibration ranges than shown and narrowed them to decrease the magnitudes of the correction terms and to improve the fit of the calibration model so that the RMS deviation between known and calculated concentrations was less than 1% relative to the average concentration. Table 1 shows the information used to design the set of standards needed for calibration and the resulting calibration ranges.

Calibration ranges were chosen according to two criteria: the expected range of concentrations in the majority of submitted samples and the ranges in which the mathematical corrections were small enough to yield desired accuracy. The second criterion required us to impose upper limits on the calibration ranges as shown in Table 1. By comparing the ranges in the table, it appears the calibration method severely limits the calibration ranges with respect to the compositions of the materials. That is indeed the case for Ca, Cu, P, S, and Zn. Fortunately, the majority of samples fit neatly into the calibration ranges. Therefore a relatively small number of samples must be diluted to be analyzed.

Dilutions are prepared by weight in mineral oil. Homogeneous mixing is assured by using a high speed rotor/stator mixer. Not all samples are miscible with mineral oil, but the shearing action of the mixer improves the miscibility of some materials. Many samples are received without composition estimates. Samples for which we have no estimates of composition are analyzed in the neat form to get estimates. Dilution factors are calculated to create samples that have concentrations of all elements within the calibration ranges. Compromises are often necessary. Our experience and consultations with our clients usually resolve any problems. If a low concentration component would be diluted below its calibration minimum, alternative analysis techniques must be used. As many as four other techniques may be required in place of XRF. Completing the analyses may take hours or even days if this happens.

Performance

Efficiency and speed are great, but what about performance in terms of accuracy and precision? The performance of the XRF additive elements method in comparison to other analysis methods will be illustrated later in the paper. At this point, the discussion will focus on quality standards and control limits derived from daily use of the method.

Two levels of control are actively applied to the method. A control sample is prepared in quantity by mixing leftovers from submitted samples and doping the mixture with other sources of any elements whose concentrations must be increased. The goal is to create an "average" sample. The initial data for using this composite material are gathered over a period of days before the previous batch is depleted. The control sample is measured before and sometimes after a batch of samples. If the method is out of control, a drift correction monitor is measured to correct for instrument instability. If the drift monitor cannot correct the situation, recalibration standards are used to recalculate the slopes and intercepts. The drift monitor is needed nearly every day. Recalibration has not been necessary except after preventative maintenance.

Table 2 illustrates the performance of the control sample over a period of one month. Control limits are set at three sigma levels and are recalculated periodically using the 20 most recent measurements. The relative uncertainties allowed by the control limits for each element are all less than 3% except for magnesium. The apparent poor performance of magnesium is to be expected due to the facts that our spectrometer is relatively insensitive for Mg and the control sample contains a very low amount of Mg. The performance of a given element could be improved but at a significant cost in terms of longer counting time. The thin polypropylene film used in the sample cups presents a practical limit on counting times, because it begins to deform after about 3 min in a 3000 W X-ray beam.

Applicable ASTM Methods

Two ASTM methods are known to be directly applicable to the XRF methods we employ. ASTM D 2622 was first published in 1967 and revised most recently in 1987. A new method, ASTM Test for Elemental Analysis of Lubricant and Additive Components—Barium, Calcium, Phosphorus, Sulfur, and Zinc—by Wavelength-Dispersive X-ray Fluorescence Spectroscopy (D 4927), was approved in 1989. Our method can be compared to both of these.

ASTM D 2622 is a simple method designed for manual treatment of X-ray spectrographic data for sulfur in four ranges between 0.001 and 5.0 wt%. We have computer-aided data reduction techniques that obviate the need to draw three separate curves manually. The calibration model in D 2622 does not correct for the presence of interferences. Our method accounts for the presence of interfering elements but not for ethanol. Quality control measurements and drift corrections are also automated.

Our method does not conform to D 2622. We have restricted our calibration range to 0.01 to 2.5 wt% using the DeJongh alphas algorithm to account for interelement influences. We do not subtract a background measurement from the peak intensity. We use di-*n*-butylsulfide and several NIST SRMs as standards, and we use mineral oil to prepare standards and samples. Since lubricating oil samples are complex, we must limit our use of D 2622 to fuel samples. Our method for additive elements is best not compared to D 2622 but rather to D 4927.

The recently approved ASTM method applies to the elements Ba, Ca, P, S, and Zn. Alternative calibration methods, internal standards or interelement correction factors, may be chosen. This method recommends calibrant compounds and 20 compositions for stan-

TABLE 2—*Control parameters for XRF method from 20 measurements of the control sample.*

Element	Average Wt%	Control[a] Limit, wt%	Relative Amount, %
Zinc	0.1725	0.0024	1.4
Calcium	0.2862	0.0043	1.5
Sulfur	0.714	0.013	1.8
Chlorine	0.1060	0.0020	1.9
Barium	0.0220	0.0006	2.5
Copper	0.0106	0.0003	2.5
Phosphorus	0.1590	0.0039	2.5
Magnesium	0.0135	0.0027	20

[a] Control limit defined as 3*s*.

dards. Our method was developed to include three additional elements. It was calibrated using calibrant solutions created from materials analyzed by other methods at Texaco as well as by NIST and Conostan.

Our XRF method is essentially equivalent to D 4927 with three exceptions. First, we have not used background corrections for Ba, S, and Zn. Sensitivities are high enough that they are not necessary. Second, we have added Cl, Cu, and Mg calibrations. These three elements are very important to our researchers. Third, and most important, we have not used the recommended calibrant sources to prepare standards. To assure ourselves and our customers that we possess an ASTM compatible method, we must acquire the octoate, sulfonate, and phosphate calibrants suggested by D 4927. If we are able to accurately analyze solutions prepared from them, we will consider our method to be compatible with D 4927.

A Coordinated Effort

Supporting Methods

At the Texaco Research Center, our elemental analysis capabilities encompass instrumental techniques such as inductively coupled argon plasma atomic emission spectroscopy (ICP-AES), flame atomic absorption spectrometry, and high temperature combustion sulfur determination. We also use classical wet chemical techniques such as an in-house gravimetric phosphorus method derived from the ASTM Test for Phosphorus in Lubricating Oils and Additives (D 1091) and Test IP 149 for Phosphorus in Lubricating Oils–Quinoline Phosphomolybdate Method. These methods are well known and will not be discussed herein except to briefly describe our modifications of them and our practices for their uses in support of X-ray fluorescence.

ICP-AES analyses of lubricants are accomplished by digesting the samples in acids (by a procedure similar to that in ASTM D 1091, Sections 10 and 11) and running them as aqueous solutions versus standards prepared from ultrapure wires and calcium carbonate. These standards are independent of any organic based standards; therefore ICP-AES analyses are completely independent of XRF analyses and provide a means to illuminate biases or corroborate results.

Flame atomic absorption spectrometry is used to perform analyses in accordance with ASTM Test for Analysis of Barium, Calcium, Magnesium, and Zinc in Unused Lubricating Oils by Atomic Absorption Spectrometry (D 4628). We have modified the method slightly by preparing solutions in aviation jet fuel (avjet). Since avjet generally has lower aromatic hydrocarbons, it burns more cleanly leaving less carbon residue on the burner head. We used potassium cyclohexanebutyrate as the source of potassium; however, we found it difficult to get into kerosene solution, even using 50 mL of 2-ethylhexanoic acid per liter of ionization suppression solution. To date, we have used this method for oils containing Ba, Ca, and Zn with performance equal to or better than the published repeatability and reproducibility statements. It has not been used for Mg analyses.

ASTM Test for Sulfur in the Analysis Sample of Coal and Coke Using High Temperature Furnace Combustion Methods (D 4239) is written for solid samples; however, it can be applied to the analysis of lubricating oils and additives by substituting liquid Standard Reference Materials of sulfur in residual fuels and using an alumina combustion aid in the boats [*2*]. ASTM Test for Sulfur in Petroleum Products (High Temperature Method) (D 1552) states that the iodometric titration method is applicable to samples boiling above 350°F and containing not less than 0.06% sulfur. Therefore it is logical to extend Method C of D 4239 (IR absorbance detection of SO_2) to the analysis of oils and additives with high boiling points, since the combustion methods are the same as D 1552.

Our extension of ASTM D 4239 to oils will hereafter be referred to as TRC4239. Using TRC4239, we have observed similar performance in terms of repeatability and reproducibility for oils analyses as compared to solids analyses. However, one caution is in order. We have observed a bias with respect to XRF analyses when working with highly overbased sulfonate additives. Use of XRF or an alternative technique is recommended for sulfur analyses in these materials.

Our current in-house gravimetric method for determining phosphorus in lubricating oils and additives is a combination of ASTM D 1091 and IP 149. The procedure uses "Quimociac Reagent" to precipitate phosphorus from dilute nitric acid solution. The gravimetric procedure is used regardless of the phosphorus content of the sample. The range of phosphorus concentrations this procedure is capable of determining is 0.005 to 25 wt% with a total relative bias of 2% or better. Absolute precision at low levels is comparable to the photometric finish of D 1091. At high levels, precision and accuracy are significantly better than D 1091. Our in-house method represents an improvement with respect to the capabilities of ASTM D 1091 or IP 149 achieved through the combination of procedures from both methods to create a single gravimetric method for all applications. Hereafter, our gravimetric method will be referred to as TRC1091.

Comparison of Results

A few Texaco materials have been analyzed by all applicable methods to compare and contrast the methods and the results. Tables 3 and 4 illustrate data from two such examples. Table 3 contains results from an extensive analysis of an additive package. The XRF measurements were made on diluted samples, and the ICP-AES samples were prepared by wet-ashing with acids. Table 4 contains results for a motor oil. No sample preparation was necessary prior to XRF analysis. The methods compared are XRF, ICP-AES, TRC1091 gravimetric phosphorus method, and TRC4239 combustion sulfur method.

Comparisons between methods are exceptionally good. All relative differences between XRF and ICP-AES are less than 3%. Likewise, the relative differences between XRF and gravimetric phosphorus results are less than 3%. The relative differences between XRF and TRC4239 sulfur results are 5%. Overall, comparisons of the 99% confidence levels to the grand averages show that uncertainties due to random errors and differences between methods are relatively small.

Accuracy is often difficult to assess for a single method. The purpose of these analyses was to compare methods that utilize different sources of standard materials, thereby exposing

TABLE 3—*Comparison of analysis results for an additive package.*

	Calcium, wt%		Copper, ppm		Phosphorus, wt%		Sulfur, wt%		Zinc, wt%	
	XRF	ICP	XRF	ICP	XRF	TRC1091	XRF	TRC4239	XRF	ICP
$\overline{X}$	2.127	2.118	591.8	587.2	0.897	0.9016	2.244	2.35	1.0182	1.0153
s	0.019	0.012	3.4	5.0	0.015	0.0016	0.030	0.04	0.0056	0.0057
n	12	8	11	7	12	10	12	8	12	8
$\overline{\overline{X}}$	2.121		589.5		0.9015		2.275		1.0171	
s_1^a	0.017		4.1		0.0112		0.033		0.0056	
s_2^a	0.011		3.1		0.0068		0.039		0.0036	

[a] s_1 is the pooled standard deviation.
[b] s_2 is the 99% confidence interval calculated for all 18 to 22 measurements of each element.

TABLE 4—*Comparison of analysis results for a motor oil.*

	Calcium, wt%		Copper, ppm		Phosphorus, wt%		Sulfur, wt%		Zinc, wt%	
	XRF	ICP	XRF	ICP	XRF	TRC1091	XRF	TRC4239	XRF	ICP
$\bar{X}$	0.219	0.215	146.1	144.3	0.133	0.1292	0.597	0.629	0.1456	0.150
s	0.002	0.005	0.9	0.3	0.003	0.0006	0.011	0.008	0.0006	0.002
n	13	11	13	5	13	16	13	8	13	11
$\bar{\bar{X}}$	0.2184		144.74		0.1293		0.6119		0.1462	
s_1[a]	0.0033		0.81		0.0018		0.0098		0.0011	
s_2[a]	0.0020		0.74		0.0013		0.0110		0.0016	

[a] s_1 is the pooled standard deviation.
[b] s_2 is the 99% confidence interval calculated for all 18 to 29 measurements of each element.

any biases inherent in a given method. We did find a small bias in our copper analyses by XRF. The bias, identified as errors in several calibration standards, was corrected.

A potential source of uncertainty in an XRF analysis is dilution of a sample in mineral oil. Each sample of the material in Table 3 was diluted by a factor of five; the material in Table 4 was not diluted. Repeatability, reproducibility, and accuracy are not significantly altered when dilution is employed; therefore a variable-speed mixer-homogenizer is an acceptable means of preparing samples diluted by weight in mineral oil. Chlorine must be held as a possible exception to this conclusion. We have seen poorer reproducibility of Cl results when analyzing diluted samples. An investigation will be undertaken in the near future.

Non-Routine Analyses

Sometimes we receive requests for nonroutine analyses. These may be of two types: requests where the sample is fully identified as containing an unusual element and those where the requester has not informed us of any change. Two examples are presented below to illustrate each of these cases and our responses to them. The first case involves the submission of barium containing samples for analysis by an XRF method that did not account for Ba without the requester or the analyst being aware of its presence. The second case involves the determination of sodium in research samples. Sodium is rarely needed but the requester was in a hurry. Since Na was not calibrated on the XRF spectrometer, ICP-AES provided Na results that were used to correct XRF results for other elements.

The sample case for barium in oils was designed to give a typical problem for analysts (i.e., incomplete information about samples) and to show how we explored and overcame the problem. By working closely together, we noticed certain samples were sent for XRF analyses (Ca, Cl, Cu, P, S, and Zn at that time) while being sent for Ba by ICP-AES. It quickly became apparent that XRF determinations of several elements were susceptible to significant errors if barium was not analyzed or at least used to correct the other elements. We decided to include Ba, and for the same reason Mg, in the XRF method.

The data that prompted this move are provided as Table 5. It shows the errors caused by ignoring barium. The table contains results from three XRF methods: the original six element method for Ca, Cl, Cu, P, S, and Zn (XRF3); the original six element version with a theoretical alpha for Ba (XRF2); and the current method with eight elements (XRF1). Table 5 also contains data from ASTM D 4628 used to corroborate the XRF calibration for Ba. The chemical values for Ba came from a well analyzed additive used to spike the 008S

TABLE 5—*Effects of Ba on analyses by XRF methods compared with ASTM D 4628.*

Sample	Chemical Wt %	U[a] Wt %	D 4628 Wt %	XRF1[c] Wt %	XRF2[d] Wt %	XRF3[e] Wt %
Barium Results						
008S0	0.000	. . .	0.000	0.000		
008S1	0.100	0.002	0.103	0.099		
008S2	0.499	0.011	0.503	0.481		
008S3	0.998	0.022	1.017	0.977		
Calcium Results						
008S0	0.167	0.006[b]	0.171	0.163	0.167	0.164
008S1	0.167		0.169	0.163	0.164	0.163
008S2	0.167		0.176	0.164	0.161	0.163
008S3	0.167		0.170	0.163	0.158	0.161
Zinc Results						
008S0	0.120	0.004	0.114	0.121	0.119	0.119
008S1	0.120		0.115	0.120	0.111	0.112
008S2	0.120		0.113	0.118	0.104	0.105
008S3	0.120		0.112	0.116	0.100	0.101
Copper Results						
008S0	0.0120	0.0004		0.0116	0.0122	0.0119
008S1	0.0121			0.0117	0.0110	0.0109
008S2	0.0124			0.0123	0.0095	0.0091
008S3	0.0129			0.0128	0.0080	0.0074
Phosphorus Results						
008S0	0.107	0.007		0.108	0.108	0.107
008S1	0.107			0.109	0.108	0.106
008S2	0.107			0.108	0.108	0.108
008S3	0.107			0.108	0.106	0.102
Sulfur Results						
008S0	0.542	0.04		0.516	0.519	0.517
008S1	0.561			0.534	0.535	0.531
008S2	0.635			0.607	0.618	0.612
008S3	0.728			0.696	0.714	0.697

[a] 99% confidence interval calculated from 8 determinations of spiking material.
[b] 99% confidence interval calculated from 20 determinations of original material.
[c] Ba fully calibrated.
[d] Ba theoretical alpha added to XRF3.
[e] Ba ignored.

samples. The chemical values for the other five elements come from the same type of thorough analysis scheme illustrated in Tables 3 and 4. Copper and sulfur were present in the barium additive; the chemical values were adjusted accordingly.

As is obvious from the XRF3 values, ignoring Ba completely when performing XRF analyses causes large errors, especially at 0.998 wt% Ba in 008S3. Zinc and copper are severely affected through absorption by Ba. As expected, the XRF3 values for Cu and Zn decrease as the Ba concentration increases. The largest effect occurs for sample 008S3 where Cu is down to 57% of the chemical value. In contrast, all XRF1 results for Ca, Zn, Cu, and P agree with the chemical values within the uncertainties listed. Sulfur results do not change from XRF3 to XRF1 and are systematically low compared to the chemical values.

Several explanations are possible for this observation. Among them are systematic differences between XRF1 and TRC4239 or incomplete correction of the Ba absorbance in XRF1.

Method XRF2 was created by adding theoretical alphas for the effects of Ba on the six elements in method XRF3 without otherwise changing the calibrations. It was expected that some improvement of the results would be observed. However, there was no significant improvement from XRF3 to XRF2. The correction for Ba must be made as a complete calibration for the effects of Ba on all elements in a method. From which it follows that one may as well calibrate for determination of the Ba itself.

Results for research samples containing 5 to 15 wt% Na illustrate an apparently successful application of results from another instrument to correct an XRF method for an element that is not determined by XRF. Table 6 compares analyses by ICP-AES and XRF of seven samples. The requester wanted quantitative results for Na, Ca, Cl, and S. Theoretical alpha coefficients for the effects of Na on Ca, Cl, and S were added to an existing XRF calibration. The theoretical alphas were calculated for samples containing 1.5 wt% Na, 0.2 wt% Ca, 0.1 wt% Cl, and 0.2 wt% S because experience indicated that the samples should be diluted by a factor of ten before the correction could be successfully applied. Calcium was analyzed by both methods to provide a check on the accuracy of the XRF results. The table shows a worst case of 8.5% relative difference between ICP-AES and XRF Ca values for sample 0215. The relative differences between the two methods are random in sign and magnitude. The average absolute relative difference with respect to ICP-AES is 4.9%. When reporting the XRF results, we assumed the accuracy for the three elements was adequately described by this comparison.

The outcome of the above experiments in matrix corrections is not entirely clear. Apparently, one may use theoretical alphas and previously determined concentrations in some cases but they must be used with care and some verification work must be done. The case of Na in sulfonates shows how well the Ca analyses worked but, in light of the example for Ba, the Cl and S results should be verified in some way. The idea of tacking a theoretical alpha for an uncalibrated element onto an existing method is not a quick fix to solve a unique analysis request. Work is necessary to ensure that each such application is correct. Future work on this topic will explore theoretical corrections for light elements such as boron, carbon, nitrogen, and oxygen.

TABLE 6—*Comparison of calcium results for sodium sulfonates. Na and Ca determined by ICP; XRF results corrected using ICP Na values with theoretical alpha.*

	ICP Results, wt %		XRF Results, wt %		
Sample	Na	Ca	Ca	Cl	S
0215	14.6	0.258	0.236	0.147	1.61
0220	13.4	0.275	0.256	0.126	1.68
0223	13.4	0.133	0.126	0.088	1.79
0228	13.8	0.145	0.147	0.086	1.68
0230	13.5	0.178	0.187	0.082	1.66
0233	14.2	0.096	0.098	0.091	1.69
0234	8.2	0.013	<0.03	0.082	4.04

•Average Relative Difference |(XRF − ICP)/ICP| = 4.9%.
•Maximum Relative Difference (Ca, #0215) = 8.5%.

Conclusion

We have presented the tools used for additive element analyses of Texaco lubricants and additive packages. Principal among them is X-ray fluorescence. It is supported and augmented by a variety of modern and classical techniques. In this way, high volumes of samples may be rapidly and accurately analyzed enabling analysts to handle the many unique materials and special requests by the best means. In our experience, it is essential that all methods be shown to agree with one another. In that way, they can be used to check each other or as alternative methods and still allow the analysts to have confidence that comparisons of analytical results by clients will be valid.

Acknowledgments

The authors wish to thank R. L. Renza, Manager, Analytical Section, Texaco Research Center, for permission to publish this work. We are also grateful to D. D. Cumming, M. W. Dilandro, and J. E. H. Macklin, who contributed to sample preparations and methods development.

References

[*1*] deJongh, W. K., *X-Ray Spectrometry,* Vol. 2, 1973, p. 151.
[*2*] Leco Corp., "SC-32 780-600 SC132 781-400 Sulfur Systems," Instruction Manual, 1987, pp. 5, Al-A5.

Johanna Y. Shay[1] *and Paul W. Woodward*[1]

Energy Dispersive X-Ray Fluorescence Spectrometric Determination of Vanadium, Nickel, and Iron in Petroleum and Petroleum Residua

REFERENCE: Shay, J. Y. and Woodward, P. W., "**Energy Dispersive X-Ray Fluorescence Spectrometric Determination of Vanadium, Nickel, and Iron in Petroleum and Petroleum Residua,**" *Modern Instrumental Methods of Elemental Analysis of Petroleum Products and Lubricants, ASTM STP 1109*. R. A. Nadkarni, Ed., American Society for Testing and Materials, Philadelphia, 1991, pp. 128–135.

ABSTRACT: Trace metals such as vanadium, nickel, and iron in fluid catalytic cracking feedstocks deposit on the catalyst and adversely affect catalyst activity and product yields. Vanadium compounds can also cause damage to furnace linings and corrosion of boiler tubes or gas turbine blades. A rapid and sensitive analytical method is needed to detect these metals in feedstocks, products, and residua. Such a method using energy dispersive X-ray fluorescence (EDXRF) spectrometry for determination of vanadium, nickel, and iron in petroleum and petroleum residua is described.

In order to eliminate matrix effects, samples containing an internal standard (cobalt) were decomposed by heating with concentrated sulfuric acid and the resulting carbonaceous residua were ashed. X-ray fluorescent spectral intensities of V, Ni, Fe, and Co were measured using established instrumental parameters. Since the Fe K_β and Co K_α lines overlapped, the slope and the intercept of the ratio of Fe K_α to Fe K_β intensities from the analysis of ash of several iron standards were used for correction of Co spectral intensities. Sample metal concentrations were determined by comparison of spectral ratios to those of ashed calibration standards. Analytical data obtained by this procedure for National Institute of Standards and Technology (NIST, formerly NBS) Standard Reference Materials (SRMs) 1618, 1634b, and 8505 were in very good agreement with the NIST certified values. The metal contents of petroleum and petroleum residua determined by this EDXRF method also agreed with those obtained by a flame atomic absorption method. This analytical approach could be developed into an ASTM standard method and the scope extended to include determination of other metals.

KEY WORDS: petroleum, metal analyses, X-ray fluorescence, vanadium, nickel, iron

Vanadium and nickel are the most abundant trace metals in petroleum. Trace metals such as vanadium, nickel, and iron in fluid catalytic cracking feedstocks deposit on the catalyst and adversely affect catalyst activity and product slate. Vanadium compounds can also cause damage to furnace linings and corrosion of boiler tubes or gas turbine blades [*1*]. A rapid and sensitive analytical method is needed to detect these metals in feedstocks, products, and residua.

An important advantage of an energy-dispersive X-ray fluorescence spectrometer is its ability to analyze simultaneously numerous elemental concentrations in a single sample.

[1] IIT Research Institute, National Institute for Petroleum and Energy Research, P.O. Box 2128, Bartlesville, OK 74005.

However, it is necessary to ash the sample to eliminate matrix effects and to concentrate elements of interest for trace metals analyses. Davis and Hoeck [*2*] developed a wet ashing method for the analysis of V, Ni, and Fe in residual fuels and feed stocks by X-ray emission spectrography. A wet ashing procedure in the presence of sulfuric acid was mentioned by Horeczy et al. [*3*]. A method for measurement of V and Ni in crude oil above 5 ppm was described by Lopp and Claypool [*4*]. Dwiggins and Dunning [*5*] reported the direct determination of Ni in oils using Co as the internal standard. Hale and King [*6*] developed a rapid method for estimating nickel at the 0.1 ppm range in refinery streams by a multichannel curved crystal X-ray fluorescence spectrometer. Two ashing methods using sulfuric acid and benzenesulfonic acid were utilized for measurement of V and Ni by Shott et al. [*7*]. Rowe and Yates [*8*] described an extension of Shott's method for determination of Cu and Fe as well as V and Ni. The absence of interelement effects between V, Ni, and Fe was demonstrated by Bergmann and co-workers [*9*]. None of the above referenced methods were performed using EDXRF. The present work with EDXRF uses a combination of the above methods plus a correction for overlap of the iron K_β and cobalt K_α lines.

Experimental Section

Instrumentation

A Princeton Gamma-Tech (PGT) Model 3800 energy dispersive X-ray fluorescence system, which consists of a System 4 analyzer interfaced to a PGT 800 XRF materials analyzer, was used for this study. The System 4 served as the multichannel analyzer and data processor, while the PGT 800 microprocessor calibrated the instrument and collected the spectral data. The PGT 800 EDXRF spectrometer uses a cryogenically cooled solid-state lithium-drifted silicon detector and a silver-anode X-ray tube.

Procedure

Calibration standards were prepared by diluting 5000 ppm Conostan standards (Conoco, Ponca City, Okla.) with Conostan base oil 75 to make stock solutions containing 200 ppm each of vanadium, nickel, iron, and cobalt. Then 20 ppm and 2 ppm calibration standards of V, Ni, and Fe were prepared by further dilutions. Finally, as shown in Table 1, stock solutions (or a blank), 4.2 g of internal standard (cobalt), and about 1 g of concentrated sulfuric acid were weighed into clean Vycor crucibles to make different ratios of metals to

TABLE 1—*Calibration standards preparation.*

Arbitrary Equivalent Concentration, ppm	Conc. of Metal Stock (V, Ni & Fe), ppm	Wt. of Metal Stock (V, Ni & Fe), g	Wt. of 200 ppm Co Stock, g	Wt. of Sulfuric Acid, g
200	200	20	4.2	1.0
100	200	10	4.2	1.0
50	200	5	4.2	1.0
20	20	20	4.2	1.0
10	20	10	4.2	1.0
5	20	5	4.2	1.0
2	2	20	4.2	1.0
1	2	10	4.2	1.0
0	base oil	0	4.2	1.0

Co concentration for preparation of calibration curves. Sulfuric acid reacted with the metals in the standards to produce metal sulfates and thus prevented them from being volatilized at the high ashing temperature. Crucibles containing the acidified standards were heated on a hot plate until all the sulfuric acid evaporated, and the mixtures were then ignited with a bunsen burner. After a dry carbonaceous material was obtained, the crucibles were placed in a programmable ashing furnace (Fisher Model 497) at 675°C until carbon-free residues were obtained.

The cooled ashes, about 10 mg, were transferred to polypropylene films (Chemplex Industries Inc.) that had been placed on top of a PGT sample cup (40 mm OD). Then the ashes were covered with a second polypropylene film, leaving the powdered ash sandwiched between the two layers of film. The X-ray fluorescent spectral intensities of vanadium, nickel, iron, and cobalt were collected once and then repeated with the sample cup rotated 180° using the following parameters:

- Run time: 100 s
- X-ray tube voltage: 40 keV
- X-ray tube current: 0.5 mA
- Filter: molybdenum
- Helium: 1.0 ft^3/h

When the molybdenum filter was used, the background in the energy region of the transition elements was substantially reduced [*10*]. In the absence of a filter, a large peak for silver from the silver-anode X-ray tube would appear.

Since the Fe K_β and Co K_α lines overlapped (Fig. 1), the slope and intercept of the ratio

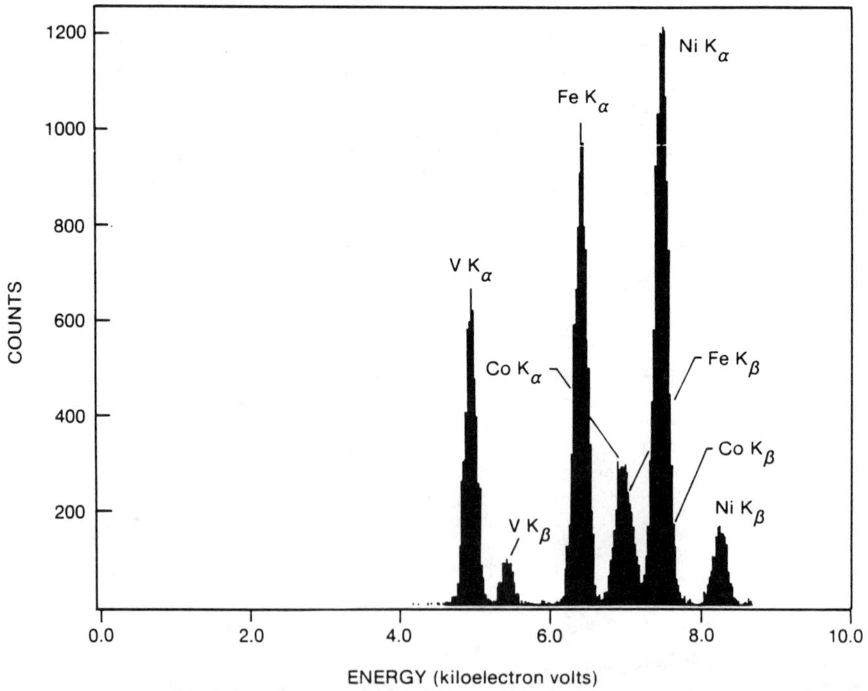

FIG. 1—*Spectrum of ashed vanadium, nickel, iron, and cobalt standards showing the overlaps of Fe* K_β *with Co* K_α, *and Co* K_β *with Ni* K_α.

of Fe K_α to Fe K_β intensities from the analysis of ash of several iron standards were used for correction of Co spectral intensities. Then the ratio of individual element counts to net counts of Co (gross counts of Co–Co counts contributed by Fe K_β line) was calculated. Calibration curves were established by using these count ratios versus the elemental concentration ratio of the standards.

The same amount of internal standard, Co, and concentrated H_2SO_4 were added to petroleum (~16 g) and petroleum residua (~8 g), then they were ashed by the same procedures as used for standards. Sample metal concentrations were determined by comparison of their spectral ratios (individual element counts/net counts of Co) to those of ashed calibration standards.

Results and Discussion

The method for wet ashing with concentrated sulfuric acid was chosen because the dry ashing procedure introduced inaccuracies from loss of volatile metalloporphyrins and the use of benzenesulfonic acid did not significantly improve results. Essentially no difference was noted between these two ashing techniques (concentrated sulfuric acid and benzenesulfonic acid) by Shott et al. [*7*]. The ashing procedure can force the detection limit much lower by concentrating the sample and eliminating the heavy background caused by scattering of X-radiation by elements of low atomic number [*11*].

Ashed standards are very stable and, because of the non-destructive nature of the EDXRF analysis, they can be reused over a long time period. When old ashed standards were analyzed as unknown samples using calibration curves established with new ashed standards, the concentration of old ashed V, Ni, and Fe standards did not change significantly after one year of storage (Table 2). Therefore it is not necessary to ash standards often, even though new calibration curves need to be established during daily operation.

A comparison of data using three different techniques employing standards for determination of metal concentrations is shown in Table 3. The method used in this study employed an internal cobalt standard to compensate for inter-element effects, thus avoiding complex calculations. The standard addition method is a good one for use when the matrixes of the sample and standard differ. However, it is very time consuming and the detection limit is high. Multi-element Conostan S-21 standards (30, 50, 100, 300, 500 ppm, and 0 ppm (blank)) were used for external standard calibration curves. The external standard method without ashing generally gave values less than 50% of those reported by the other methods because of matrix difference between sample and standard.

The results of three standards (SRM 1618, 1634b, and 8505) obtained from this procedure agree very well with NIST certified values (Table 4). The metal contents of petroleum and petroleum residua determined by this EDXRF method generally agreed with those obtained by a flame AA method, although nickel and iron contents obtained by AA were, in general, slightly lower than those from EDXRF (Tables 4 and 5). The vanadium content of 87100 crude determined by AA is apparently low based on results for the 87100 residue.

To check the effect of sample orientation in the X-ray path, EDXRF spectral intensities (counts) of both standards and samples were measured twice. The second intensities were obtained after rotating the sample cup about 180°. The effect on analytical results due to sample cup rotation is shown in Table 6. The difference may not be highly significant, but precision is improved by using an average of counts obtained both before and after cup rotation.

The higher concentration of iron standard produced the higher Fe K_β intensity. If the overlap of the Fe K_β and Co K_α was not considered, a higher Co K_α value would have been obtained. Therefore the ratio of metals to Co for the higher concentrations would have

TABLE 2—*Aging effect on ashed standards.*

Standard No.	Year	Ashed Standards, ppm Vanadium	Nickel	Iron
St1	0	200	200	200
	1	199	204	201
St2	0	93.8	98.7	99.9
	1	92.7	98.9	101.0
St3	0	48.0	48.8	50.2
	1	48.1	49.3	50.1
St4	0	21.4	20.2	20.5
	1	21.0	20.0	20.3
St5	0	10.0	9.9	10.3
	1	10.1	10.0	10.3
St6	0	4.9	4.9	5.1
	1	4.8	4.9	5.1
St7	0	2.0	1.9	2.0
	1	1.9	1.8	2.1
St8	0	1.3	0.8	1.3
	1	1.3	0.7	1.4

TABLE 3—*Comparison of X-ray fluorescence results using different methods of quantitation with those obtained from round-robin study.*

	External Standard			Standard Addition			Internal Standard			Round Robin[a]		
Crude ID	V, ppm	Ni, ppm	Fe, ppm	V, ppm	Ni, ppm	Fe, ppm	V, ppm	Ni, ppm	Fe, ppm	V, ppm	Ni, ppm	Fe, ppm
Maya	138	29	ND	278	55	ND	275	57	5.0	260 ± 26	53 ± 6.0	4.0 ± 2.9
California	97	45	5.0	202	93	NA	199	80	79	198 ± 24	77 ± 8.0	70 ± 17

ND = not detected, NA = not available.

[a] Round robin was conducted by the Phillips Petroleum Company in 1984. Various analytical methods were used: AA, ICP, XRF.

TABLE 4—*Comparison of SRM data from XRF*[a] *and AA methods with NIST certified values.*

SRM No.	Vanadium, ppm XRF	AA	NIST	Nickel, ppm XRF	AA	NIST	Iron, ppm XRF	AA	NIST
8505	391 ± 19	378	390 ± 10	55.6 ± 2.9	47.5	...	9.8 ± 0.5	5.4	...
1618	422 ± 11	397	423 ± 3	78.3 ± 5.0	73.9	75.2 ± 0.4	6.5 ± 1.0	5.5	...
1634b	51.0 ± 0.5	50.9	55.4 ± 1.1	27.4 ± 1.1	23.1	28.0 ± 2.0	32.1 ± 1.9	22.5	31.6 ± 2.0

[a] Data were obtained from replicates: 6, 9, and 10 for SRM 8508, 1618, 1634b, respectively.

TABLE 5—*Comparison of metal contents obtained by XRF and AA techniques for crudes and petroleum residua (>565.6°C).*

Sample No.	Vanadium, ppm				Nickel, ppm				Iron, ppm			
	Measured		Mass Balance[a]		Measured		Mass Balance[a]		Measured		Mass Balance[a]	
	XRF	AA	XRF	AA	XRF	AA	XRF	AA	XRF	AA	XRF	AA
8731 crude	0.3	<0.1	0.4	0.5	0.8	0.6	1.1	0.8	1.3	0.5	1.6	0.9
87100 crude	51.1	40.5	47.6	53.6	14.0	11.2	14.2	11.7	6.0	2.4	4.0	3.3
8930 crude	2.6	3.2	3.2	4.1	3.0	2.6	3.2	2.5	4.2	2.1	3.4	2.5
8943 crude	2.9	3.3	3.2	3.6	2.3	2.0	2.5	2.1	1.7	0.9	1.9	1.3
8979 crude	0.7	1.6	0.9	1.1	<0.1	<0.3	0.2	<0.2	0.8	0.5	2.0	1.3
8731 resid	5.2	6.0			12.3	9.8			18.7	10.4		
87100 resid	244	275			72.8	59.8			20.4	16.8		
8930 resid	28.9	36.4			28.8	22.3			30.4	22.0		
8943 resid	29.2	33.6			22.8	19.4			17.7	11.6		
8979 resid	11.0	13.4			2.0	<2.0			23.9	15.7		

[a] Mass balance = Resid fraction of whole crude × percent metal content of resid, assumes that all metals remain in the resid during distillation.

significantly decreased (Table 7). Because all three NIST standards (1618, 1634b, and 8508) have low iron content, the ratio of metals to Co did not change significantly between net counts and gross counts of Co used. However, 1618 and 8508 have high vanadium content; thus the high V calibration standards were needed for calculation. When gross counts of Co were used, higher V results were obtained because of the big difference in percent decrease of metals/Co ratio between the calibration standards and NIST samples (Table 7). While the effect from overlap of the Fe K_β and the Co K_α (Fig. 1) was significant for higher concentrations of metals, the overlap did not strongly affect lower concentrations of metals (Table 7) because the intensity produced by the trace Fe K_β was not significant.

A standard with low cobalt content was used for the internal standard, because high cobalt content may cause the cobalt K_β line to interfere with measurement of the nickel K_α line (Fig. 1). Table 8 shows that the determination of nickle content of samples was not affected by a low concentration of cobalt.

TABLE 6—*Comparison of X-ray fluorescence spectral intensities before and after sample cup rotation.*

STD No.	Before Rotation			After Rotation		
	Ratios of Counts (Conc., ppm)			Ratios of Counts (Conc., ppm)		
	V/Co (V)	Ni/Co (Ni)	Fe/Co (Fe)	V/Co (V)	Ni/Co (Ni)	Fe/Co (Fe)
STD 1	2.154(198)	4.583 (204)	3.918(208)	2.219(204)	4.555(202)	3.882(206)
STD 2	1.113(93.6)	2.579(98.5)	2.010(99.0)	1.118(94.0)	2.592(98.9)	2.048(101)
STD 3	0.559(47.2)	1.350(48.5)	1.018(49.4)	0.578(48.8)	1.368(49.1)	1.052(51.0)
STD 4	0.209(21.8)	0.619(20.4)	0.425(20.6)	0.202(21.0)	0.608(20.0)	0.420(20.4)
STD 5	0.110(9.6)	0.332(9.9)	0.210(10.2)	0.114(10.0)	0.333(9.9)	0.214(10.4)

TABLE 7—*Effect of the overlap of iron* K_β *and cobalt* K_α.

	Cobalt Counts Used for Calculation								
	Net Counts			Gross Counts			% Decrease		
NIST No.	V/Co	Ni/Co	Fe/Co	V/Co	Ni/Co	Fe/Co	V/Co	Ni/Co	Fe/Co
200 ppm STD	2.513	5.493	4.285	1.911	4.141	3.230	24.0	24.6	24.6
100 ppm STD	1.210	2.673	2.035	1.065	2.353	1.792	12.0	12.0	11.9
50 ppm STD	0.632	1.391	1.040	0.591	1.301	0.972	6.5	6.5	6.5
20 ppm STD	0.241	0.622	0.412	0.234	0.606	0.401	2.9	2.6	2.7
10 ppm STD	0.125	0.334	0.206	0.123	0.329	0.204	1.6	1.5	1.0
5 ppm STD	0.062	0.181	0.106	0.062	0.180	0.106	0.0	0.6	0.0
2 ppm STD	0.030	0.106	0.038	0.030	0.106	0.038	0.0	0.0	0.0
1 ppm STD	0.018	0.067	0.023	0.018	0.067	0.023	0.0	0.0	0.0
0 ppm (Blank)	0.005	0.048	0.007	0.005	0.048	0.007	0.0	0.0	0.0
NIST 1618	1.388	0.651	0.036	1.387	0.651	0.036	0.1	0.0	0.0
NIST 1634b	0.386	0.636	0.489	0.375	0.617	0.474	2.8	3.0	3.1
NIST 8505	1.477	0.541	0.064	1.477	0.541	0.064	0.0	0.0	0.0
	V, ppm	Ni, ppm	Fe, ppm	V, ppm	Ni, ppm	Fe, ppm			
NIST 1618	427	79.1	6.4	522	84.8	6.5			
NIST 1634b	50.0	27.7	31.0	50.2	27.8	31.6			
NIST 8505	392	56.5	9.9	486	58.9	10.0			

Conclusion

A method that utilizes energy dispersive X-ray fluorescence spectrometry has been established for determination of metals content in a wide variety of petroleum samples. The procedures were developed specifically for determination of nickel, vanadium, and iron, but can easily be extended for determination of many other elements.

Pre-treatment with sulfuric acid, ashing of samples prior to analysis, and use of low concentration cobalt internal standard eliminates matrix effects for low detection limits with high precision and accuracy. The prepared ashes are stable and can be kept for long periods

TABLE 8—*Effect of low cobalt content on the determined concentration of nickel.*

	Nickel Counts Used for Calculation	
	Net Counts[a]	Gross Counts[a]
Sample No.	Nickel Content, ppm	Nickel Content, ppm
NIST 1634b	28.0	28.0
NIST 1618	79.1	79.1
87116 crude	9.0	9.0
8767 crude	9.6	9.7
8749 crude	5.5	5.5
8624 crude	1.0	1.0

[a] Net nickel counts = Gross nickel counts − Nickel counts contributed by cobalt K_β line.

of time, and because of the nondestructive nature of EDXRF they can be reused many times for further investigations.

An important aspect of the method is the development of a procedure for correction of the overlap of iron K_β and cobalt K_α lines was developed.

Acknowledgments

The authors are grateful to the USDOE Office of Strategic Petroleum Reserve (OSPR) and the USDOE Bartlesville Project Office (BPO) for financial support of this work. Appreciation is also expressed to Dr. Ray Anderson of NIPER and Mr. Harry Giles of OSPR for technical and administrative assistance, and to Mrs. Judy Green, NIPER, for atomic absorption analyses.

References

[1] Johns, M. C. K. and Hardy, R. L., *Industrial and Engineering Chemistry,* Vol. 44, 1952, p. 2615.
[2] Davis, E. N. and Hoeck, B. C., *Analytical Chemistry,* Vol. 27, 1955, p. 1880.
[3] Horeczy, J. T., Hill, B. N., Walters, A. E., Schultze, H. G., and Bonner, W. H., *Analytical Chemistry,* Vol. 27, 1955, p. 1899.
[4] Lopp, V. R. and Claypool, C. G., *Advances in X-Ray Analysis,* W. M. Mueller, Ed., Plenum Press, New York, Vol. 3, 1960, p. 131.
[5] Dwiggins, C. W. and Dunning, H. N., *Analytical Chemistry,* Vol. 31, 1959, p. 1040.
[6] Hale, C. C. and King, W. H., Jr., *Analytical Chemistry,* Vol. 33, 1961, p. 74.
[7] Shott, J. E., Jr., Garland, T. J., and Clark, R. O., *Analytical Chemistry,* Vol. 33, 1961, p. 506.
[8] Rowe, W. A. and Yates, K. P., *Analytical Chemistry,* Vol. 35, 1963, p. 368.
[9] Bergmann, J. G., Ehrhardt, C. H., Grantelli, L. and Janik, J. L., *Analytical Chemistry,* Vol. 39, 1967, p. 1258.
[10] Bilbrey, D. B., Leland, D. J., Leyden, D. E., Wobrauschek, P., and Aiginger, H., *X-Ray Spectrometry,* Vol. 16, 1987, p. 161.
[11] Kang, C. C., Keel, E. W., and Solomon, E., *Analytical Chemistry,* Vol. 32, 1960, p. 221.

Bradner D. Wheeler[1]

Elemental Analysis of Crude and Lubricating Oils by Energy Dispersive X-Ray Fluorescence

REFERENCE: Wheeler, B. D., **"Elemental Analysis of Crude and Lubricating Oils by Energy Dispersive X-Ray Fluorescence,"** *Modern Instrumental Methods of Elemental Analysis of Petroleum Products and Lubricants, ASTM STP 1109,* R. A. Nadkarni, Ed., American Society for Testing and Materials, Philadelphia, 1991, pp. 136–147.

ABSTRACT: Most aspects of petroleum research, exploration, and refining require elemental analysis from trace to minor elemental concentrations. Numerous elements ranging from atomic numbers 15 to 92 can be found in many crudes. Elements such as vanadium, iron, nickel, copper, and arsenic can have deleterious effects on the catalysts utilized in the catalytic cracking process. Other elements purposely introduced as additives into finished lubricating oils must also be measured and controlled. Typically elements such as phosphorus, sulphur, calcium, zinc, and molybdenum are in the range of 0.00 to 5.00 wt%.

In addition to crude oils and finished lubricating oils, engine and turbine bearing wear analysis are often performed by X-ray fluorescence. In these cases, typical bearing and piston ring elements such as copper, chromium, zinc, lead, tin, and silver are usually quantified.

The techniques for preparing samples of crude oil, finished lubricating oils, and used engine oils are basically identical. The sample is transferred into a polypropylene sample cup with a 6.3 μm polypropylene support film. Errors in volume or sample depth can produce significant intensity deviations for higher energy X-rays. The error for lead, for example, is approximately 1.5% relative per 0.1 mL error at volumes of 3 mL. Minimum detectable limits or several tenths of parts per million for atomic numbers 30 to 92 from a range of 5 to 800 parts per million can be expected.

KEY WORDS: X-ray fluorescence, minimum detectable limit, parts per million, atomic number, intensity, crude oil, lubricating oil, absorption, enhancement, peak overlap

Many aspects of petroleum research, exploration, and refining require the analysis of a variety of elements in the trace to minor concentration ranges. Vanadium, nickel, and sulphur are the most deleterious contaminating elements in crude oil and their geochemistry is well known. Elements such as arsenic, iron, vanadium, nickel, and copper have adverse effects on catalysts used in the cracking process, while other elements such as magnesium, phosphorus, sulphur, chlorine, zinc, molybdenum, and barium must be controlled in the final products. In addition, engine and turbine bearing wear analysis must be quantified to determine the degree of wear. This analysis can be performed in a similar manner as the elemental analysis of lubricating oils and greases utilizing X-ray fluorescence.

Wavelength X-ray fluorescence has been widely used in process control for approximately 30 years; in the past 15 years energy dispersive X-ray fluorescence has become more widely utilized and accepted. Analysis of many petroleum products has been accomplished by

[1] LINK Analytical, Redwood City, CA 94065.

classical gravimetric and volumetric techniques, inductively coupled plasma, optical emission, and atomic absorption. Although these analytical procedures are capable of providing viable results, they can be prone to errors in weighing and dilutions.

Instrumentation

Analysis of these materials was performed utilizing an energy dispersive X-ray spectrometer operated under the instrumental parameters shown in

• Assayed element:	Mg, P, Cl, K, Ca, Zn	Mo
• Anode:	Ag	Ag
• Anode voltage:	15 kV	45 kV
• Anode current:	200 μA	150 μA
• Energy range:	0-10 keV	0-20 keV
• Atmosphere:	He	He
• Filter:	none	Rh
• Counting time:	300 s	40 s

Sample Preparation

Sample preparation for analysis of petroleum products is a relatively easy task, since all that is required are polypropylene sample cells with a thin 6.3 μm polypropylene support film. The sample is usually transferred to the sample cell in a manner to ensure that the volume or depth of the sample remains constant. Errors in sample volume or depth can introduce significant errors in the measured intensities of the higher energy X-rays as can be observed from the intensity versus sample depth curve (Fig. 1) [*1*]. Also, the selection of the support film must take into consideration that the film is strong enough to support the sample, resistant to attack by the sample, and have good transmittance of X-rays for the elements of interest. Examination of Fig. 2 [*2*] suggests that the best choice of film would be 6.3 μm polypropylene.

Considering that the sample depth or volume and the thickness of the polypropylene support film are constant, the minimum detectable concentration for the elements on the periodic table from $Z = 11$ to 92 would be as shown in Fig. 3. Generation of the data for Fig. 3 was obtained by analysis of Conostan standard sample C-21 [*3*].

Quantification of Results

X-ray fluorescence can be an excellent technique for the quantitative analysis of petroleum. The technique can be quite precise and accurate, but due to various interelement effects the intensity and the apparent concentration may not reflect the true values. A simple expression designating the possible variations could be

$$C_i = K \cdot I_i \cdot M \cdot S \tag{1}$$

where

C_i = concentration of analyte element *i*,
K = geometrical configuration of the instrument including take-off angle, incidence angle, detector resolution, etc.,
I_i = measured intensity of analyte element *i*,

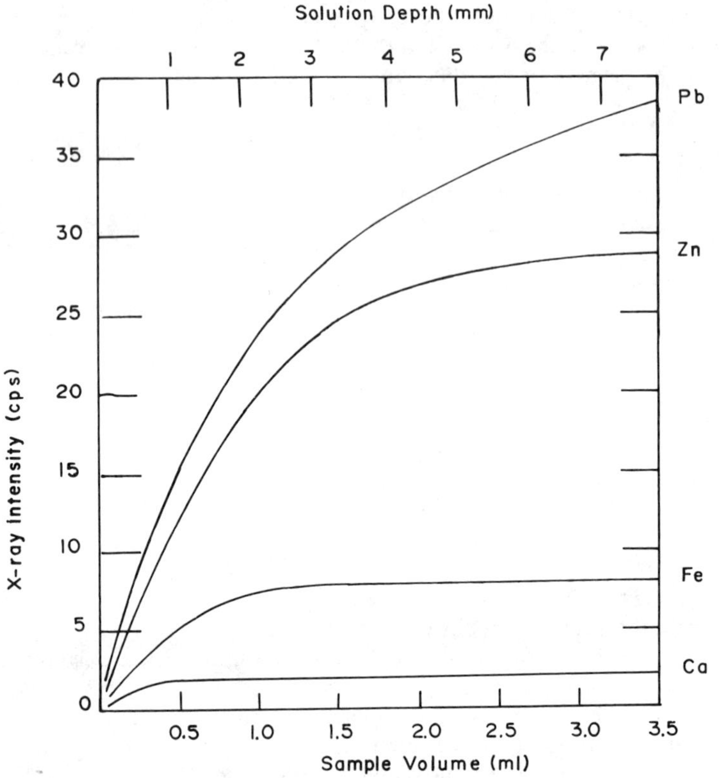

FIG. 1—*Graphical illustration of the effect of solution volume and depth versus X-ray counting rate.*

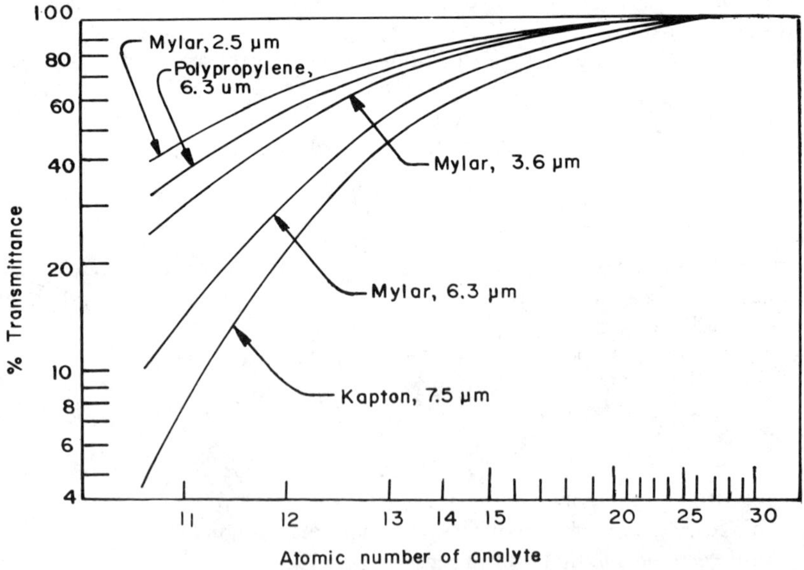

FIG. 2—*Transmittance of X-rays through liquid support films versus atomic number of analyte.*

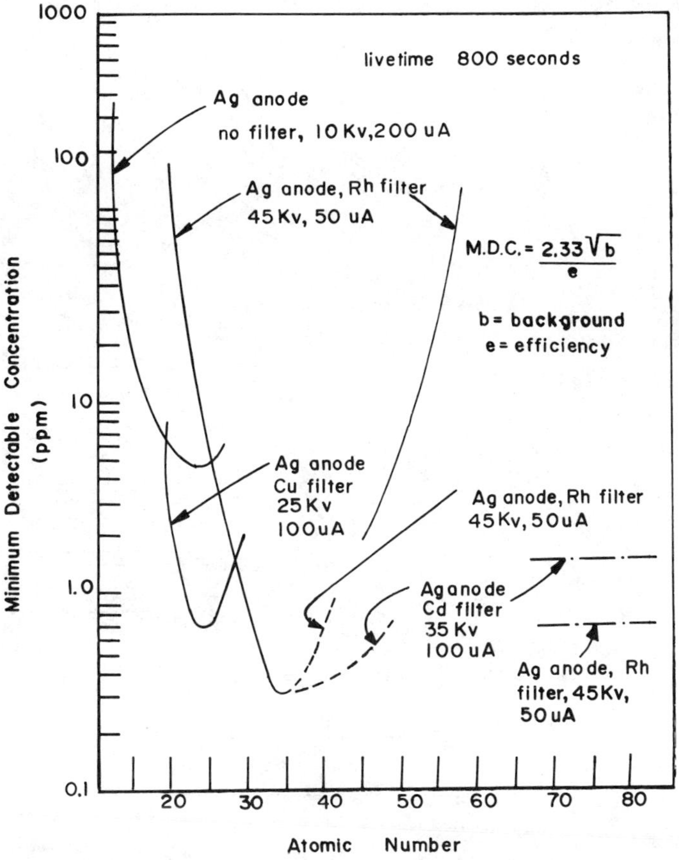

FIG. 3—*Minimum detectable concentration versus atomic number for a low Z matrix (dashed portions of curves are estimated extrapolations).*

M = mathematical model utilized for interelement influence corrections, and
S = sample preparation techniques such as depth or volume of sample, transmittance of support film, and homogeneity of the sample.

X-ray fluorescence, like other instrumental methods of analysis, is not free of spectral line interferences of interelement effects. Figure 4 is a graph of the mass absorption coefficient versus energy in kiloelectron volts along with the absorption edges of the associated analytical lines. Examination of Fig. 4 reveals several cases of absorption and enhancement plus severe spectral overlaps. The most pronounced case is the analysis of sulphur in the presence of phosphorus and molybdenum. In this example, sulphur is greatly enhanced by the presence of molybdenum and strongly absorbed by phosphorus due to the phosphorus K edge being on the low energy side of the sulphur K alpha line. Numerous mathematical models are available [*4-13*], which range from fundamental parameters to empirically derived coefficients. Since most of these models rely on fundamental parameters or are predicated on determining influence coefficients based on concentrations, complete knowledge of the composition for the standards is imperative. This work, however, was based on empirically derived coefficients based on the measured intensities [*14,15*].

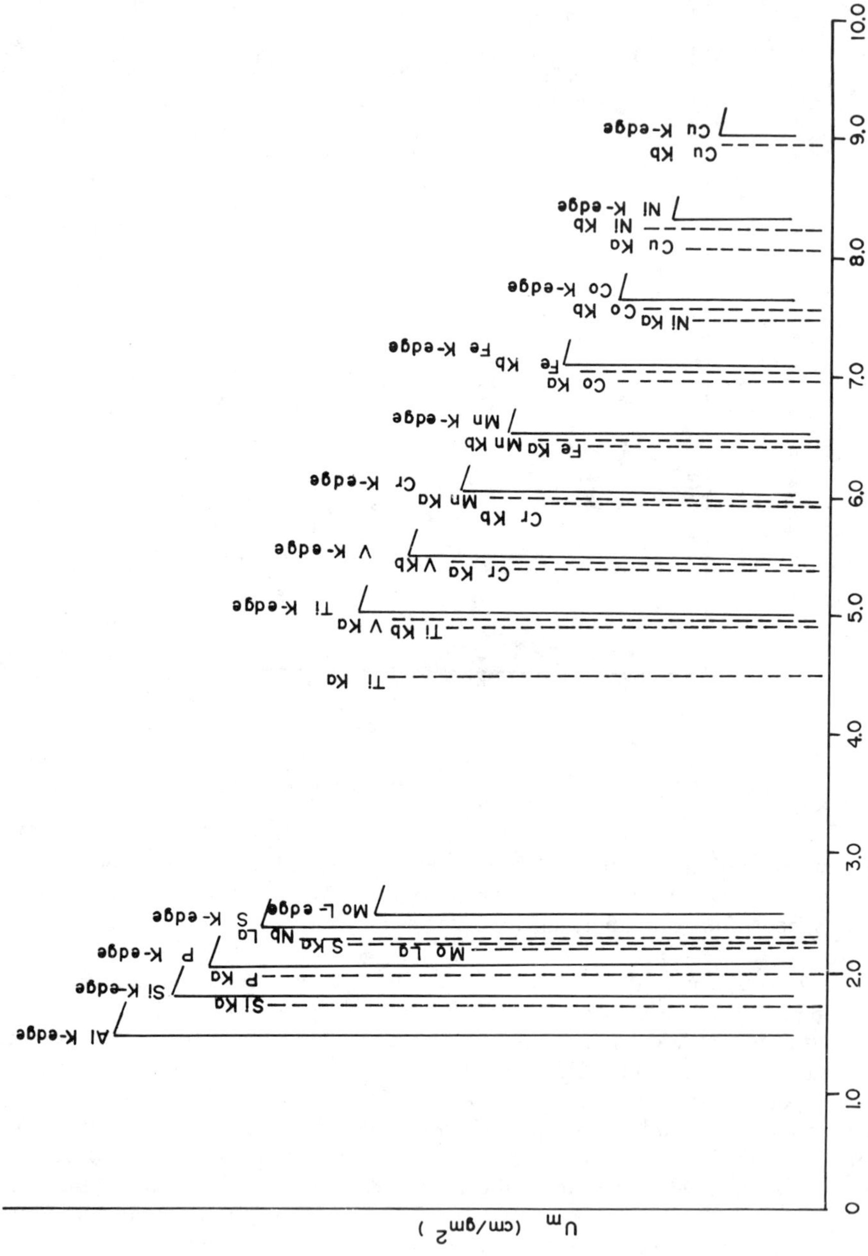

FIG. 4—*Mass absorption coefficient versus energy (in keV).*

The generalized equation, an adaptation of Beer's Law, used in this study is

$$C_i = a + bI_i \exp - m(I_j - I_{jj}) \tag{2}$$

where

C_i = concentration of element i,
a = x-intercept of calibration curve,
b = slope of calibration curve,
I_i = intensity of analyte element i,
I_j = intensity of interfering element j,
I_{jj} = average intensity of interfering element j for all standards, and
m = influence coefficient derived by regression analysis.

In the specific case of the analysis of sulphur in the presence of molybdenum and phosphorus the equation would appear as

$$C_s = a + B[ISK\alpha + Mol\alpha - (R\ Mol\alpha/Mok\alpha \cdot IMok\alpha]\exp - m(I_p - I_{pj})) \tag{3}$$

where

C_s = concentration of sulphur in standard,
a = x-intercept of calibration curve,
b = slope of calibration curve,
$ISk\alpha + Mol\alpha$ = measured intensity of sulphur k-alpha and molybdenum l-alpha lines,
$R\ Mol\alpha/Mok\alpha$ = ratio of intensity of molybdenum l-alpha line to intensity of molybdenum k-alpha line,
$IMok\alpha$ = measured intensity of $Mok\alpha$ line in sample,
I_p = measured intensity of phosphorus,
I_{pj} = average intensity of phosphorus k-alpha line in all standards, and
m = influence coefficient derived by regression analysis.

For this model to be mathematically correct, the number of standards must be $n + 6$ where n is the number of corrections made on any single element. In this example two corrections for molybdenum and phosphorus are being made on sulphur. As a result the

TABLE 1—*Interelement corrections for X-ray fluorescence analysis of petroleum.*

Analyte Element	Interfering Element							
	Mg	P	S	Cl	K	Ca	Zn	Mo
Mg	...	...	...	...	...	...	...	...
P	...	...	...	...	...	...	...	...
S	...	a	...	e	...	...	...	o
Cl	...	a	a	...	...	e	...	...
K	...	...	a	a	...	e	...	...
Ca	...	...	a	a	a	...	e	...
Zn	...	...	...	...	...	a	...	a
Mo	...	...	...	...	...	...	...	...

NOTE:
a = absorption
e = enhancement
o = spectral overlap

TABLE 2—*Analytical results of oil analysis by energy dispersive X-ray fluorescence.*

Sample I.D.	P		S		Cl		Ca		Zn		Mg		K		Mo	
	list	calc	list	calc	list	calc	list	calc	list	calc	list	calc	list	calc	list	calc
L-1	0.1028	0.1029	0.4819	0.4968	-	-	0.0555	0.0574	0.1094	0.1131	-	-	-	-	-	-
L-2	0.1340	0.1364	0.5180	0.5169	-	-	0.1560	0.1581	0.1450	0.1437	0.0460	0.0480	-	-	-	-
L-3	0.0869	0.0868	0.4221	0.4453	-	-	0.4510	0.4535	0.0987	0.1006	0.0022	0.0030	-	-	0.0199	0.0200
L-4	0.1320	0.1329	0.4960	0.4908	-	-	0.1900	0.2002	0.1640	0.1641	-	-	-	-	-	-
L-5	0.1290	0.1292	0.6120	0.6103	-	-	0.1900	0.1967	0.1580	0.1611	-	-	-	-	-	-
L-6	0.0894	0.0869	0.3800	0.3929	-	-	0.2320	0.2291	0.0986	0.1012	0.0017	0.0025	0.0063	0.0065	0.0203	0.0202
L-7	0.0356	0.0363	0.4850	0.4878	-	-	0.8580	0.8703	0.0420	0.0435	0.0030	0.0025	-	-	-	-
L-8	0.0377	0.0379	0.5760	0.5881	-	-	1.2200	1.2480	0.0443	0.0456	0.0043	0.0050	-	-	-	-
L-9	0.0348	0.0362	0.7320	0.6932	-	-	1.2400	1.2147	0.0440	0.0413	0.0041	0.0050	-	-	-	-
L-10	0.0344	0.0341	0.5400	0.5558	-	-	1.4600	1.4515	0.0382	0.0386	0.0083	0.0090	-	-	-	-
L-11	0.1070	0.1107	0.4280	0.4321	-	-	0.0019	0.0037	0.1270	0.1267	-	-	-	-	-	-
L-12	0.1460	0.1401	0.5910	0.5645	-	-	0.3690	0.3547	0.1800	0.1733	0.0035	0.0030	0.0042	0.0039	0.1117	0.1120
L-13	0.0690	0.0698	0.3720	0.3736	-	-	0.0974	0.0945	0.0854	0.0831	-	-	-	-	-	-
LC-1	-	-	-	-	2.1300	2.1681	-	-	-	-	-	-	-	-	-	-
LC-2	-	-	-	-	1.0500	0.9983	-	-	-	-	-	-	-	-	-	-
LC-3	-	-	-	-	0.0600	0.0630	-	-	-	-	-	-	-	-	-	-
L-14	0.0271	0.0248	0.3950	0.4057	-	-	0.0053	0.0058	0.0315	0.0300	-	-	-	-	-	-
L-15	0.0429	0.0444	0.1403	0.1466	-	-	0.0352	0.0375	0.0520	0.0533	-	-	-	-	-	-
L-16	0.0613	0.0610	0.1600	0.1630	-	-	0.0146	0.0142	0.0742	0.0744	-	-	-	-	-	-
L-17	0.0337	0.0340	0.3933	0.3930	-	-	0.0073	0.0080	0.0420	0.0417	-	-	-	-	-	-
L-18	-	-	0.1350	0.1346	-	-	0.1920	0.1950	-	-	-	-	-	-	-	-
L-19	0.0157	0.0161	0.3750	0.3741	-	-	0.0004	0.0002	-	-	-	-	-	-	-	-
LC-4	0.0075	0.0080	2.0200	1.9605	0.0043	0.0075	0.0043	0.0044	0.0053	0.0050	-	-	-	-	-	-
LC-5	-	-	0.2360	0.2471	0.1030	0.1203	-	-	-	-	-	-	-	-	-	-
LC-6	0.0239	0.0243	0.5360	0.5360	1.2500	1.2985	-	-	0.0231	0.0229	-	-	0.0031	0.0029	-	-
LC-7	-	-	0.4380	0.4569	0.9130	0.8674	0.0003	0.0002	0.0002	0.0003	-	-	-	-	-	-
LC-8	-	-	-	-	0.5560	0.5846	-	-	-	-	-	-	-	-	-	-
LC-9	-	-	-	-	3.0500	3.0538	-	-	-	-	-	-	-	-	-	-
LC-10	-	-	-	-	3.0500	3.0536	-	-	-	-	-	-	-	-	-	-

Notes: list = listed or certified value
calc = calculated value by regression analysis
- = value not listed
All values expressed as weight percent

minimum number of standards required would be eight. This is only one example of the interelement corrections required. The complete correction procedure is listed in Table 1.

The parameters *a*, *b*, and *m* in Eq 2 are all calculated by regression analysis with the necessity of only one known value, the analyte concentration. The concentration of element *j* need not be known so long as the intensity of I_j can be measured. In this particular expression, if the rare circumstance should happen that element *j* had no effect (absorption of enhancement) on the intensity of *i*, then the expression would default to a simple linear equation since the exponential term would be equal to one. If, however, *j* had an effect, the term for *m* would assume a value reflecting the magnitude of the correction with either a positive or a negative sign. A negative sign would indicate *j* as an enhancer on *i*, while a positive sign would indicate *j* as an absorber of *i*.

Discussion and Results

Most phases of petroleum research, exploration, refining, and production of the final product require the analysis of a variety of elements in the trace to minor concentrations. Numerous elements such as phosphorus, sulphur, calcium, zinc, and molybdenum are purposely introduced as additives in the final product, while other elements such as arsenic, iron, vanadium, and chromium in crude oil can have adverse effects on the catalyst utilized in the cracking process. Consequently, to achieve maximum efficiency in the refining and blending process the elements must be quantified and controlled.

Since the analyses of crude, blended, and motor oils by X-ray fluorescence are similar, this study was primarily on finished motor oil products. The mathematical expression used for the interelement corrections is an intensity model, an adaptation of Beer's Law, in which the coefficients were derived empirically based on the standards listed in Table 2. Since the results are based on intensity corrections, it was not necessary to have knowledge of the composition of the interfering elements.

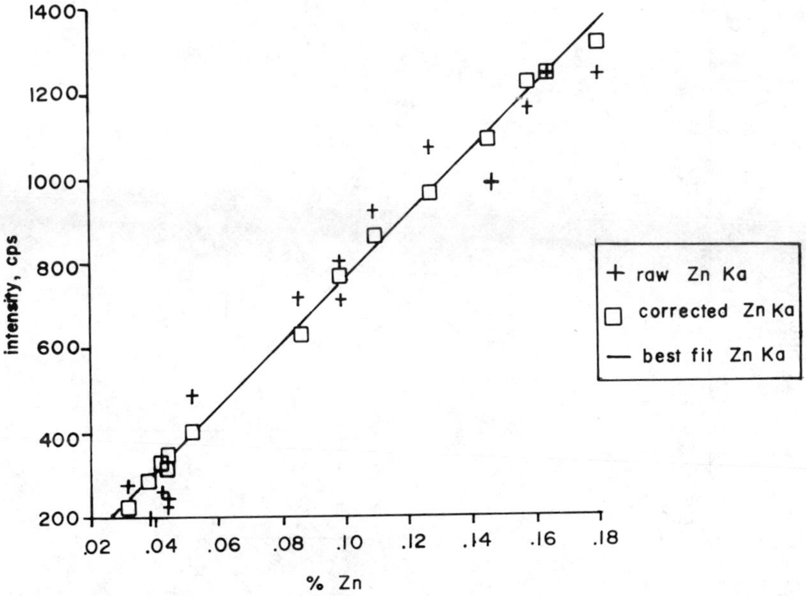

FIG. 5—*Calibration curve for zinc in motor oil.*

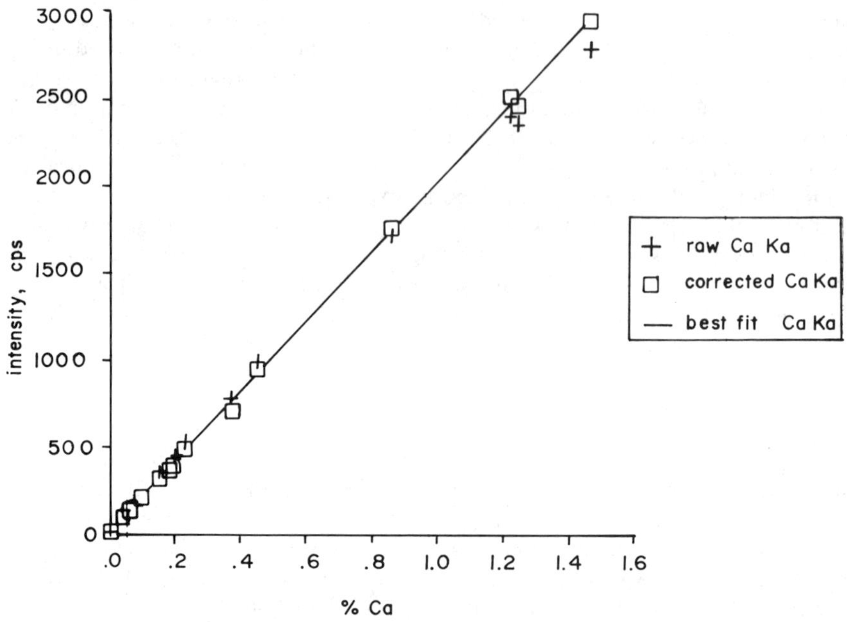

FIG. 6—*Calibration curve for calcium in motor oil.*

Table 2 represents a compilation of results based on a series of standards and utilization of interelement corrections derived through Eq 2. Figures 5 to 9 show typical calibration curves for zinc, calcium, sulphur, phosphorus, and chlorine in the finished products. Figure 10 is a typical spectra of a common blended motor oil prior to use.

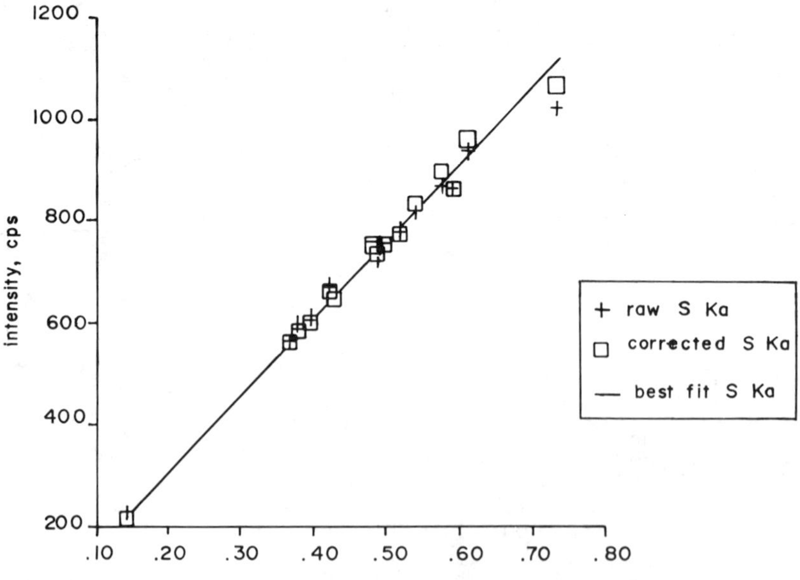

FIG. 7—*Calibration curve for sulphur in motor oil.*

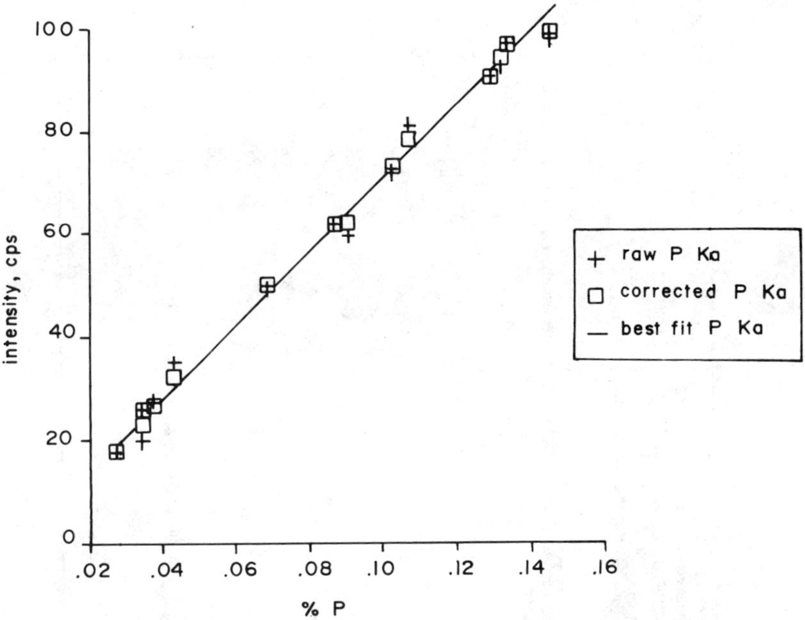

FIG. 8—*Calibration curve for phosphorus in motor oil.*

Conclusions

Process control of the refining and blending process demands precise and accurate chemical analysis of crude stock and finished product. In addition, monitoring of engine performance and bearing wear also requires elemental analysis equally as stringent. The technique of X-

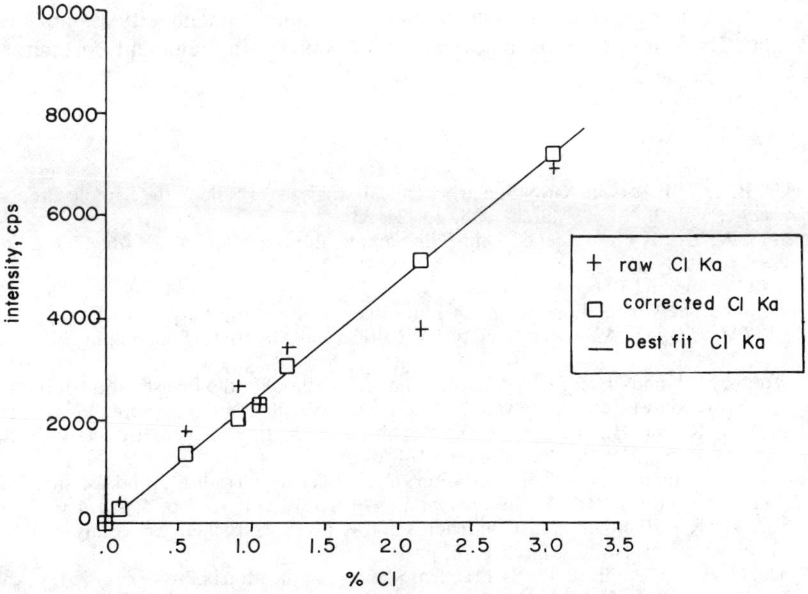

FIG. 9—*Calibration curve for chlorine in petroleum products.*

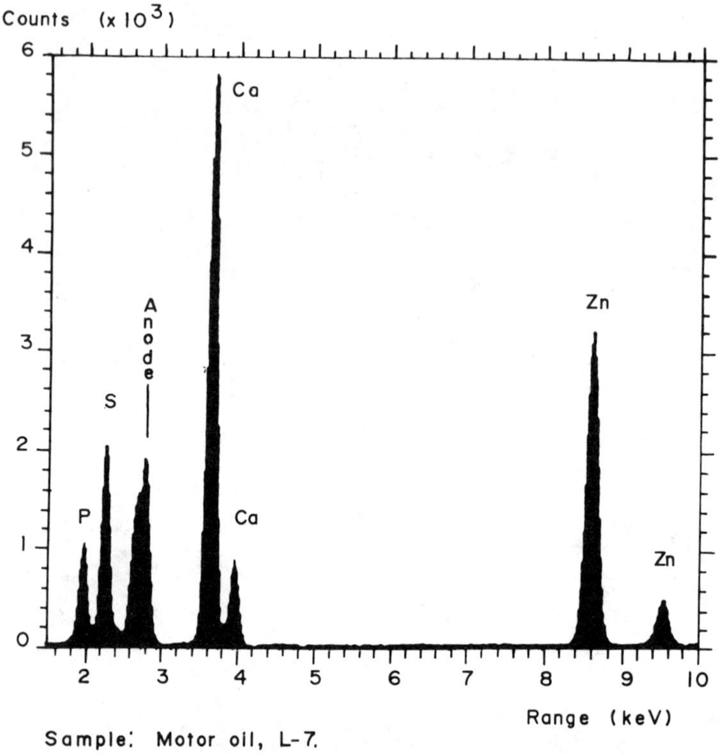

FIG. 10—*Spectra of motor oil sample L-7.*

ray fluorescence can provide the analyst with precise and accurate results. These results, however, depend on proper excitation conditions, sample homogeneity, sample volume, sample support film, an adequate mathematical model for interelement corrections, and accuracy of the standards.

References

[1] Wheeler, B. D., "Errors in X-ray Spectrochemical Analysis as Related to Sample Preparation," *Spectroscopy,* Vol. 3, No. 3, March 1988, pp. 24–33.

[2] Solazzi, M. J., "X-ray Fluorescence Thin-Film Sample Support Materials," *American Laboratory,* Nov. 1985, pp. 124–131.

[3] Conoco Oil Company, Ponca City, Okla.

[4] Sherman, J., "The Correlations Between Fluorescence X-ray Intensity and Chemical Composition," *ASTM STP 157,* American Society for Testing and Materials, Philadelphia, 1954, pp. 27–33.

[5] Lucas-Tooth, M. J. and Price, B. J., "A Mathematical Method for the Investigation of Interelement Effects in X-ray Fluorescence Analysis," *Metallurgica,* Vol. 64, No. 383, Sept. 1954.

[6] Raspberry, S. R. and Heinrick, K. F. J., "Calibration for Interelement Effects in X.R.F.A.," *Analytical Chemistry,* Vol. 46, No. 1, Jan. 1974, p. 81.

[7] Claisse, F. and Quintin, M., "Generalization of the Lachance-Traill Method for the Correction of the Matrix Effect in X.R.F.A.," *Canadian Spectroscopy,* Vol. 12, No. 4, April 1967, p. 129.

[8] Lachance, G. R., "Fundamental Coefficients for X-ray Spectrochemical Analysis," *Canadian Spectroscopy,* Vol. 15, No. 2, March 1970, p. 64.

[9] Lachance, G. R. and Traill, R. J., "A Practical Solution to the Matrix Problem in X-ray Analysis," *Canadian Spectroscopy,* Vol. 11, No. 2–3, March–May 1966.

[10] Criss, J. W. and Birks, L. S., "Calculation Methods for Fluorescent X-ray Spectrometry—Empirical Coefficients vs. Fundamental Parameters," *Analytical Chemistry,* Vol. 40, No. 7, June 1968, pp. 1080–1086.

[11] Criss, J. W., Birks, L. S., and Gilfrich, J. V., "Versatile X-ray Analysis Program Combining Fundamental Parameters and Empirical Coefficients," *Analytical Chemistry,* Vol. 50, No. 1, Jan. 1987, pp. 33–37.

[12] Rousseau, R. M. and Bouchard, M., "Fundamental Algorithm Between Concentration and Intensity in XRF Analysis," *X-ray Spectrometry,* Vol. 15, 1968, pp. 207–215.

[13] Kalnicky, D. J., "Analysis of Petroleum Products Using a Modified Alphas Approach," *American Laboratory,* Nov. 1985.

[14] Wheeler, B. D., "Proceedings of the Third Forum on the Geology of Industrial Minerals," Special Distribution 34, University of Kansas, Lawrence, 1967, pp. 72–82.

[15] Hasler, M. R. and Kemp, J. W., "Suggested Practices for Spectrochemical Computations," ASTM Committee E-2, Report E-2, SM 2-3, Philadelphia, 1957, pp. 72–82.

Gwen G. Ohlson[1] *and Yoshi Takahashi*[1]

Improved Performance for Trace Halide Determination in Hydrocarbons

REFERENCE: Ohlson, G. G. and Takahashi, Y., "**Improved Performance for Trace Halide Determination in Hydrocarbons,**" *Modern Instrumental Methods of Elemental Analysis of Petroleum Products and Lubricants, ASTM STP 1109,* R. A. Nadkarni, Ed., American Society for Testing and Materials, Philadelphia, 1991, pp. 148–154.

ABSTRACT: Microcoulometry has been widely used to measure halides in hydrocarbons. While microcoulometric analysis is well-established, some types of hydrocarbons (i.e., crude oils, heavy fuel oils, hydraulic fluids, and others) require special attention due to their physical and chemical properties. In the past, slow combustion with higher temperatures (1000°C or greater) was recommended; however, sluggish response due to refractory components, the presence of heavy metals, or the rapid aging of the quartz combustion tube under such conditions has made data interpretation difficult at times.

Analysis of heavy hydrocarbons was improved by changing the combustion tube design and adding cupric oxide wire to the combustion outlet zone. With this scheme, it is not necessary to use high furnace outlet temperatures. Combustion efficiency and system response improve, while the effects of deposits and corrosion are greatly reduced. Crude and #6 fuel oils, phosphate ester fluid, and base lube oil additives were tested under these conditions, yielding better recovery and improved precision for some by a factor of three or greater.

A post-combustion scrubber developed for solvent analysis was evaluated for refractory hydrocarbon analysis. It had little effect on the detection limit, but could be used to remove deposits.

KEY WORDS: #6 fuel oil, base lube oil additive, chlorine, combustion, crude oil, halides, microcoulometry, phosphate ester fluid, refractory hydrocarbons, titration, total halides

Trace halide impurities, primarily chlorides, in hydrocarbons are known to cause corrosion problems and to poison catalysts during processing steps. Microcoulometry has been widely used to measure total halides. It is specific, sensitive over a wide dynamic range, and is suitable for a broad range of hydrocarbon types. It is quantitative for chlorine, with partial detection of bromine and iodine, and no detection of fluorine.

For many samples, complete oxidation and reliable analysis is achieved by using an automated boat inlet and higher furnace temperatures (800 to 1000°C). While requiring these and even higher temperatures to oxidize, very heavy refractory types (such as crude oil, base lube additives, and phosphate ester fluids) have other physical and chemical properties which make reliable trace analysis sometimes difficult. Modifying the combustion scheme eliminates the effects of these properties and yields improved response and more reliable data.

[1] Dohrmann Division, Rosemount Analytical, Santa Clara, CA 95054.

Experimental Procedure

The analyzer used in this study to determine total halides is the MCTS-120 Automated Microcoulometric Titrating System, manufactured by Dohrmann Division of Rosemount Analytical. The backbone of this system is a high-temperature combustion furnace and a halide-specific microcoulometric titration cell.

The sample is oxidized to its oxidized elemental form: CO_2, H_2O, X_2, HX, HOX, NO_x, SO_x, and so forth. These gases pass into a silver titration cell, where only the halides (all chlorine, HX forms of bromine and iodine, no fluorine) react to form silver halide: $X^- + Ag^+ \rightarrow AgX$. The silver depletion in the cell from the halide reaction is sensed potentiometrically by the coulometer and replaced by sending a current through a silver electrode: $Ag^\circ \rightarrow Ag^+ + e^-$. By coulometry, this measured current is directly proportional to the halide that was in the sample.

MCTS-120 Operating Conditions

- Sample Inlet: Automatic boat, 60 s vaporization time, gas switching between Ports 1 and 3, 120 s after start
- Gas Flows: Port 1–65 cm^3/min oxygen
  Port 2–135 cm^3/min oxygen
  Port 3–5 cm^3/min helium
  Port 4–45 cm^3/min helium (across boat race)
- Combustion Tube: GT-120 design for refractory hydrocarbons (Fig. 1)

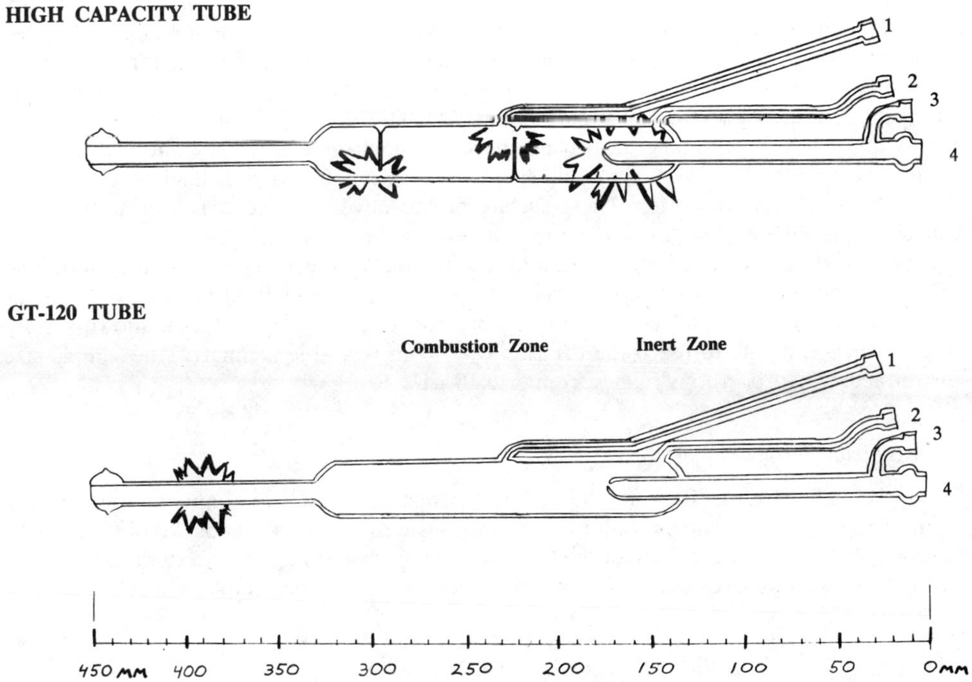

FIG. 1—*Combustion tube designs.*

- Temperature: Variable (700 to 1050°C); furnace has two independently controlled zones
- Exit Tube Packing: Cupric oxide (CuO) wire

Sample Preparation

- Crude Oil and #6 Heavy Fuel Oil: Diluted in tetrahydrofuran (THF) 1/20 and 1/10, respectively; 5 μL injections
- Phosphate Ester (Hydraulic Fluid): Analyzed neat; 5 μL injections
- Base Lube Oil Additive: Diluted 1/10 in toluene; 10 μL injections

Dilutants are all reagent grade unless otherwise noted.

Results and Discussion

Refractory hydrocarbons are hard to combust and can contain heavy metals or alkaline additives which plate onto or etch the quartz surface of the combustion tube. The halide ion that does get generated tends to interact physically with these deposits or corroded areas and is released slowly, often too slowly to generate a reliable response.

Changing the Combustion Tube Design

The high-capacity style combustion tube, originally developed for the MCTS-120, is very effective at completely oxidizing large volumes. It is wider and shorter than previous designs with baffles to promote mixing and broaden the residence time, resulting in greater capacity for many sample types. However, for refractory hydrocarbons containing metals or alkalis, the increased mixing becomes a disadvantage. Etching and surface deposits form rapidly at the inert zone (Fig. 1). With successive injections, less and less of the vaporized halide enters the combustion zone; instead, the vaporized halide "hangs-up" on the etched surfaces. Increasing the furnace temperature will improve combustion efficiency and reduce coke deposits; however, it will also increase the surface effects of metals and alkalis, often making the problem worse, not better. These surface effects develop rapidly, rendering the high-capacity style tube impractical for routine analysis of these sample types.

By removing the baffles from the tube, the surface effects disappear from the inert zone (GT-120 tube, Fig. 1); we have also seen this with similar unbaffled designs. The metals and alkalis now tend to collect in the disposable exit tube in the cool region, allowing more of the oxidized halide to reach the titration cell. With this tube geometry, the sample size for refractory hydrocarbons ranges from 5 to 10 μL.

Cupric Oxide Wire

Cupric oxide wire was first investigated by Matsuzaki et al. [*1*] to eliminate interferences from nitrogen, sulfur, and phosphorous. Cupric oxide (CuO) wire is packed into the exit tube which then sits at the edge of the GT-120 tube hot zone (Fig. 2). We studied the effect of the CuO wire on overall performance and precision for three sample types.

Cupric Oxide Wire As an Oxidation Promoter

Figure 3 demonstrates improved combustion for some heavy hydrocarbons using cupric oxide. Crude and #6 fuel oils are very refractory, traditionally requiring high furnace tem-

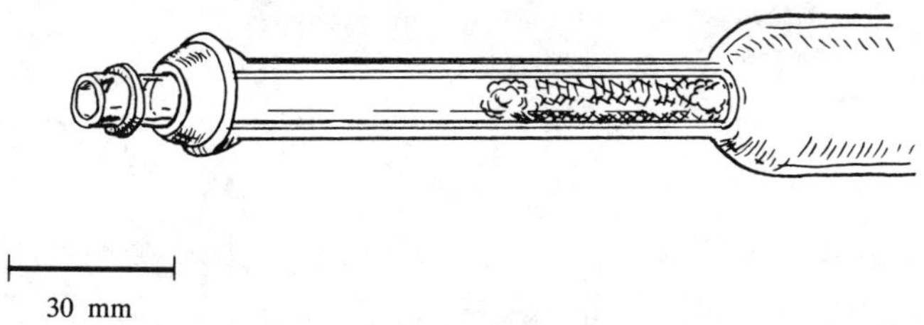

FIG. 2—*Cupric oxide in exit tube.*

peratures. The peak shapes are much more defined, particularly the #6 fuel. Also important is the difference in outlet temperatures: the cupric oxide packing improves the combustion at a lower operating temperature, which reduces the activity of the heavy metals these samples may contain.

Cupric Oxide Wire to Remove Deposits

Halide ion sticks to deposits as it travels from the combustion zone to the titration cell. This is due to mechanical "hang-up" or chemical interaction and causes sluggish response, tailing, and a shifting or noisy baseline. In addition to converting coke deposits to their oxidized gas form, cupric oxide removes metal and alkaline deposits from the gas stream, freeing the halide ion to be titrated. One important deposit that cupric oxide effectively removes is phosphorous which, as Matsuzaki et al. [*1*] noted, attaches to the cupric oxide, forming a green, cupric pyrophosphate.

An example of the effect of phosphorous and how CuO eliminates it is shown in Fig. 4. Hydraulic fluid or phosphate ester fluid contains a significant amount of phosphorous. A vapor at high temperatures (1000 to 1050°C), the phosphorous becomes sticky as it condenses in the cool zone of the exit tube. With each injection, more phosphorous collects, causing more tailing, baseline shifting, and greater error (see No CuO trace, Fig. 4). Cupric oxide removes the phosphorous from the gas stream before it can condense, and does so at a less severe temperature (700°C), allowing the halide ion to travel to the detector unimpeded. Tailing and baseline shifts disappear, resulting in improved precision and more reliable performance.

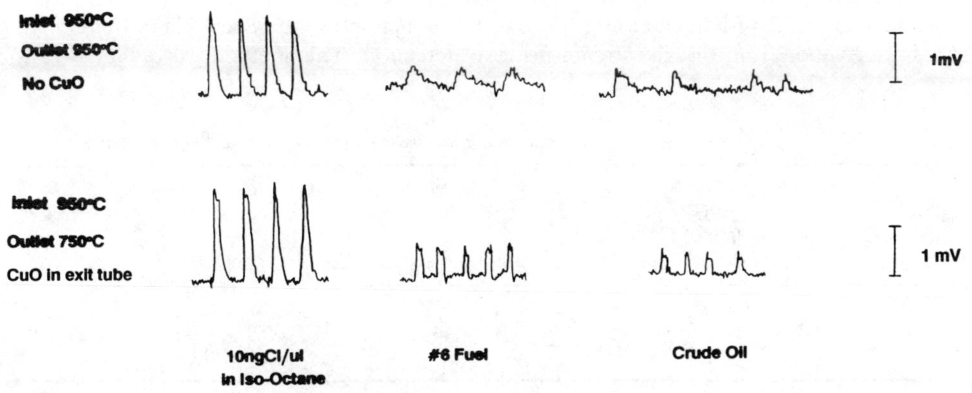

FIG. 3—*Improved combustion of heavy fuel oils.*

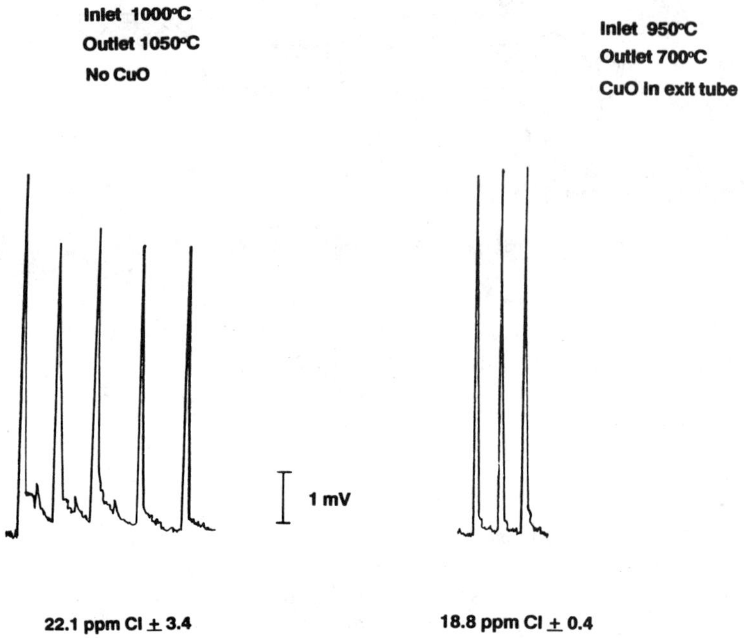

FIG. 4—*Improved hydraulic fluid analysis.*

Base Lube Oil Additive Analysis

Base lube oil additive has all the properties challenging to trace halide analysis: the halide level is low and of a refractory nature, while the material itself is very hard to combust and can contain as much as 1% of several alkaline earths, making it corrosive to the tube. Again, packing the exit tube with cupric oxide improves the combustion, allows a lower furnace temperature which reduces the surface effects, and removes alkaline deposits. The improved precision (by a factor of 3) for three base lube oil additives is shown in Table 1.

Further Improvements

The effect of water vapor on the titration cell has been demonstrated during light hydrocarbon analysis [2]. Ladrach et al. [2] showed a reduction in the system blank by removing the water vapor from the gas stream prior to detection. This method was adapted to the

TABLE 1—*Results of base lube oil additive analysis for three samples diluted in toluene.*[a]

No CuO Inlet 950°C Outlet 950°C ($ppmX^-$)	Sample	**CuO in Exit Tube** Inlet 950°C Outlet 750°C ($ppmX^-$)
0.20 ± 0.35	A	0.41 ± 0.11
1.21 ± 0.39	B	0.89 ± 0.06
1.25 ± 0.29	C	0.86 ± 0.09

[a] $n = 3$.

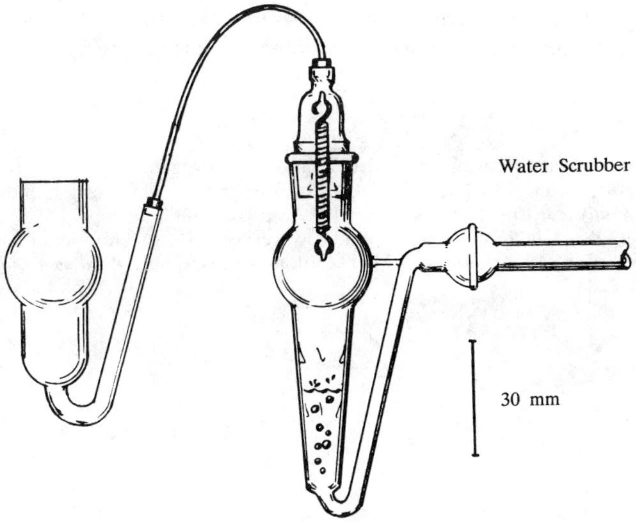

FIG. 5—*A new water scrubber.*

MCTS-120/Boat Inlet configuration to determine if removing the water vapor would improve performance for refractory samples as well.

The scrubber (Fig. 5) is filled with 3 mL of 90% (vol/vol) concentrated sulfuric acid (H_2SO_4) and placed between the combustion tube and the titration cell. Results from solvent blanks with and without the water scrubber were generated using the boat inlet/low sample volume combination for refractory samples. For comparison, the same solvent was analyzed both ways by using the syringe/high sample volume combination used for analysis of light hydrocarbons (40 uL direct syringe injection volume, 700°C inlet/800°C outlet furnace temperatures, low electrolyte level in titration cell).

While using the water scrubber did lower the blank under solvent analysis conditions, as previously reported, it had little effect on boat analysis (Table 2). Moreover, baseline noise did not diminish nor did the precision change. However, the scrubber effectively removed coke deposits from the gas stream and could benefit both solvent and refractory analyses for that purpose.

Conclusion

Using the GT-120 combustion tube and cupric oxide plus lower outlet temperatures will benefit the routine day-in/day-out analysis of heavy hydrocarbons. Reduced corrosion to

TABLE 2—*Effect of water scrubber on solvent blank.*[a]

	No Scubber (ppmX$^-$)	With Scrubber (ppmX$^-$)
10 μL using boat inlet	1.5	1.1
40 μL direct syringe injection	1.3	0.6

[a] $n = 3$; precision = ±0.1.

the tube means lower replacement costs; removing deposits means less titration cell maintenance; and improved combustion and precision means greater reliability in the numbers.

References

[*1*] Matsuzaki, A. and Koyano, K., "Elimination of Nitrogen, Sulfur, and Phosphorous Interferences in Determination of Chlorine in Petroleum Products by Combustion and Microcoulometric Titration," *Analytical Chemistry*, Vol. 52, 1980, pp. 146-149.

[*2*] Ladrach, W., Van De Craats, F., and Gouverneur, P., "Determination of Traces of Chlorine in Organic Liquids: A Combustion-Microcoulometric Method," *Analytica Chimica Acta,* Vol. 50, 1970, pp. 219-228.